CAMBRIDGE LIBRARY COLLECTION
Books of enduring scholarly value

Women's Writing
The later twentieth century saw a huge wave of academic interest in women's writing, which led to the rediscovery of neglected works from a wide range of genres, periods and languages. Many books that were immensely popular and influential in their own day are now studied again, both for their own sake and for what they reveal about the social, political and cultural conditions of their time. A pioneering resource in this area is Orlando: Women's Writing in the British Isles from the Beginnings to the Present (http://orlando.cambridge.org), which provides entries on authors' lives and writing careers, contextual material, timelines, sets of internal links, and bibliographies. Its editors have made a major contribution to the selection of the works reissued in this series within the Cambridge Library Collection, which focuses on non-fiction publications by women on a wide range of subjects from astronomy to biography, music to political economy, and education to prison reform.

The Book of Sun-Dials
Margaret Gatty (1809–73) was an English writer of popular science best known for her researches on sundials and British seaweeds. After marrying the Rev. Alfred Gatty in 1839, she moved to Ecclesfield, Yorkshire, where she pursued her literary and scientific studies. This volume, first published in 1872, contains detailed descriptions of various styles of sundials, many taken from Gatty's own collection. Over 350 sundials from across Britain and Europe are described (with their mottoes provided and translated where necessary), and each sundial's location is noted in this work, which was one of the first popular books on the subject. Examples included range from portable sundials to early Saxon sundials, as well as the more familiar church sundials. This volume is one of Gatty's best known works, and remains a valuable reference for the various types of sundials and the variations and similarities in their mottoes. For more information on this author, see http://orlando.cambridge.org/public/svPeople?person_id=gattma

Cambridge University Press has long been a pioneer in the reissuing of out-of-print titles from its own backlist, producing digital reprints of books that are still sought after by scholars and students but could not be reprinted economically using traditional technology. The Cambridge Library Collection extends this activity to a wider range of books which are still of importance to researchers and professionals, either for the source material they contain, or as landmarks in the history of their academic discipline.

Drawing from the world-renowned collections in the Cambridge University Library, and guided by the advice of experts in each subject area, Cambridge University Press is using state-of-the-art scanning machines in its own Printing House to capture the content of each book selected for inclusion. The files are processed to give a consistently clear, crisp image, and the books finished to the high quality standard for which the Press is recognised around the world. The latest print-on-demand technology ensures that the books will remain available indefinitely, and that orders for single or multiple copies can quickly be supplied.

The Cambridge Library Collection will bring back to life books of enduring scholarly value (including out-of-copyright works originally issued by other publishers) across a wide range of disciplines in the humanities and social sciences and in science and technology.

The Book of Sun-Dials

MARGARET GATTY

CAMBRIDGE UNIVERSITY PRESS

Cambridge, New York, Melbourne, Madrid, Cape Town, Singapore,
São Paolo, Delhi, Dubai, Tokyo, Mexico City

Published in the United States of America by Cambridge University Press, New York

www.cambridge.org
Information on this title: www.cambridge.org/9781108020978

© in this compilation Cambridge University Press 2010

This edition first published 1872
This digitally printed version 2010

ISBN 978-1-108-02097-8 Paperback

This book reproduces the text of the original edition. The content and language reflect
the beliefs, practices and terminology of their time, and have not been updated.

Cambridge University Press wishes to make clear that the book, unless originally published
by Cambridge, is not being republished by, in association or collaboration with, or
with the endorsement or approval of, the original publisher or its successors in title.

A SHORTENED MORTUARY CROSS
In Over Peover Churchyard, Cheshire.
CITO PEDE PRÆTERIT ÆTAS 1679. Nº 30

THE

BOOK OF SUN-DIALS.

COLLECTED BY

MRS. ALFRED GATTY,

AUTHOR OF "PARABLES FROM NATURE,"

ETC.

LONDON:
BELL AND DALDY, YORK STREET,
COVENT GARDEN.
1872.

CHISWICK PRESS:—PRINTED BY WHITTINGHAM AND WILKINS,
TOOKS COURT, CHANCERY LANE.

TO

THE DEAR HUSBAND,

TO WHOM I AM INDEBTED FOR THE BEST HAPPINESS OF

THE HOURS OF EARTHLY LIFE,

AND WITH WHOM I HOPE TO SHARE THE EXISTENCE IN WHICH

TIME SHALL BE NO MORE,

I Dedicate this Volume,

IN THE COMPILATION OF WHICH HE HAS TAKEN SO GREAT

A PART AND INTEREST.

M. G.

PREFACE.

IF any one should open these pages, expecting to find in them an astronomically scientific account of sundials, from their first simple origin to the complicated and even confused perfection at which they arrived, just before they were superseded by clocks, in the beginning of the eighteenth century, he will be disappointed.

Some years ago, when we were contemplating this publication, we rejoiced at receiving the following advice from a wise and accomplished friend, James Nasmyth, Esq.—" If I might presume to say so, I think you will do well to keep quite clear of any astronomical treatment of the subject. If I apprehend aright, your object is the poetry and moral of dials, under their varied treatment in several ages. This is the romantic part of it, the other is the dry one; and has been done many times already." If we had ever doubted the wisdom of this counsel, our hesitation would have been swept away by the sight, which through the kindness of his daughter we have been permitted to enjoy, of the models and works on Dialling of the late Rev. W. Hewson, M.A., vicar of Goatland, Yorkshire

PREFACE.

—the labour of eight years, and still requiring as long an apprenticeship to understand them. They remain a monument of his indefatigable zeal, and remind us that the undergraduates of Cambridge used to say that only Sir Isaac Newton could explain the dial which he himself had erected on Queen's College.

On the subject of this volume so much has been already said in the Introductory Chapter, which might naturally have formed a part of the Preface, that little remains to be added, beyond thanking those who have assisted in making the collection.

Among these there is one dear young friend, without whom it is probable that the work would never have appeared—Miss Eleanor Lloyd. To her the reader is indebted for by far the greater number of the continental mottoes, and for much of the pleasant notices which accompany them, as well as for general unwearied enthusiasm in her researches. Being an artist, too, she has adopted the habit which we ourselves had pursued for so many years, and made sketches of all the dials she saw, both at home and abroad.

Thanks are also due to numerous correspondents in the pages of "Notes and Queries," who, ever since we first communicated with that periodical on this subject, under the name of *Hermes*, have from time to time sent word of dials worthy of notice. A recent appeal in the columns of the "Guardian" has likewise led to some interesting contributions. Of private friends who have most kindly given their help, the list of names would be endless; and we can only hope that they will accept this assurance of how much we feel obliged to one and all.

There is, however, one more name, which must most certainly

be put on record. We have particularized a friend, but for whose contributions during the last ten years this work might probably not have appeared; it would be dishonest to omit the name of another, without whose help it could not have appeared—the Rev. Alfred Gatty, D.D. To him are wholly due the introduction and arrangement of whatever classical or antiquarian lore is to be found in the ensuing pages, as well as the judicious selection from materials which had largely accumulated during so long a period of time.

Margaret Gatty.

INTRODUCTORY.

THERE is no human discovery more ancient, or more interesting, than that of the Sun-dial: so ancient that the exquisite essayist, Charles Lamb, says, " Adam could scarcely have missed it in Paradise;" and so interesting that we may be sure that man's first want, after supplying the cravings of hunger, would be to invent some instrument by which he could measure the day-time into portions, to be allotted to his several avocations.

" Please, sir, what's o'clock?" is the child's enquiry, as he "tents" his mother's cow in the lane pastures; and the hardy backwoodsman, hewing out a settlement for himself in the primeval forest, leans on his axe, and looks to the sun's position in the heavens for information how soon he may retire to his hut for food and sleep. Time is a blank if we cannot mark the stages of its progress; and it has been found that the human mind is incapable of sustaining itself against the burden of solitary confinement in a dark room, where you can take no note of time. The great Creator, who made the sun to rule the day and the moon and the stars to govern the night, has adapted our nature to these intermitting changes, and implanted in us an immediate desire to count how, drop by drop, or grain by grain, time and life are passing away.

INTRODUCTORY.

Edgar Poe sings, in melancholy strain, as he stands in imagination on the sea-shore—

> "I hold within my hand
> Grains of the golden sand;
> How few, yet how they creep
> Through my fingers to the deep,
> While I weep!"

The first notion of dissecting time would of course be suggested by a tree, or a pole stuck in the soil, the shadow of which moving from west to east as the sun rose or declined in the sky, would lead men to indicate by strokes on the ground the gradual progression of the hours during which the daylight lasted. Further observation would discover that if the pole slanted so as to point to the north star, and run parallel with the earth's axis, a sun-dial was constructed that would measure the day. But the fixing of a complete instrument, varying in its lines and numbers, according to the locality, and whether horizontally or vertically placed, would be a matter of progressive astronomical and mathematical calculation, which only the scientific could accomplish, long after the rude art of uncivilized man had discovered the means of ascertaining midday, and dividing into spaces the morning and afternoon.

Herodotus writing 445 B.C. says, that "it was from the Babylonians that the Greeks learned concerning the pole, the gnomon and the twelve parts of the day." (B. ii. cap. 109.) These twelve parts however, would always differ in length according to the season, except at the equinox, because the ancients always reckoned their day from sun-rise to sun-set. The word "hour" therefore, as they used it, must be regarded as an uncertain space of time, until it was accurately defined by astronomical investigation.

The Jewish Scriptures, our oldest literature, give us no clear information as to how time was reckoned in the ancient world. "The evening and the morning were the first day" (Gen. i. 5) is the earliest description of a period of time whose duration we cannot surely estimate. A week is also thus defined: "On the seventh day God ended his work which he had made, and he rested on the seventh day from all his work which he had made." (Gen. ii. 2.) Farther on in the Jewish history we find the day divided into four parts, and the night into three watches. In Nehemiah ix. 3, we read, "They stood up in their place, and read in the book of the law of the Lord their God one-fourth part of the day; and another fourth part they confessed, and worshipped the Lord their God." This mode of computation appears to have lasted until our Saviour's time. In His parable of a householder hiring servants, He describes him as going out at the third, sixth and ninth hours to engage additional labourers, and afterwards at the eleventh hour before the day closed. (Matthew xx. 1—8.) The night was divided by the Jews into three watches. The "beginning of the middle watch" is spoken of in Judges vii. 19. This reckoning also lasted to the time of Christ: "Blessed are those servants whom the Lord when He cometh shall find watching . . . and if He shall come in the second watch, or come in the third watch." (Luke xii. 37, 38.)

In short, the accurate measurement of time is a comparatively late invention. What the ancients effected for this purpose was the result of a close observation of the heavenly bodies, whereby they made rude computations according to the amount of their knowledge. The exact hour of the day which the totally unlearned wayfarer now ascertains by drawing his silver watch from his fob, the Chaldean star-gazers 2,000 or 3,000 years ago deduced, as best they could, from a constant study of those hieroglyphic lights which rule day and night alternately, and whose motions are now more accurately known and noted than the arrival and departure of railway trains in the columns of Bradshaw.

But our business is with Sun-dials, and the first on historical record is that of Ahaz, who reigned over Judah 742 B.C. It has been observed that the Babylonians or Chaldeans were the first people who seem to have divided time by any systematic mechanical contrivance. A lucid atmosphere is favourable to celestial contemplation, of which the people of the East have always fully availed themselves; and even now those countries most abound in Sun-dials which have the clearest skies. The Rev. S. C. Malan thus speaks of a recent visit to Ur of the Chaldees, and the landscape of serene beauty presented to him on the site of Rebekah's well, "as the shadows of the grass and of the low shrubs around the well lengthened and grew dim, and the sun sank below the horizon, the women left in small groups; the shepherds followed them, and I was left in this vast solitude, yet not alone; the bright evening star in the glowing sky to westward seemed to point to the promised land, as when Abraham took it for his guide."

From this people of Chaldæa, these star-searchers of the old world, we may conclude that Ahaz got his notion of dialling, and we read in the history of the unfortunate reign of this king a possible, nay a likely cause of his introduction of Babylonish customs. Being pressed in war by the kings of Israel and Syria, Ahaz sought alliance and rescue from Tiglath Pileser, king of Assyria, who indeed released him in his emergency, but made him pay heavy tribute, and conform his worship to that of the Assyrians. "The altars at the top of the upper chamber of Ahaz" (2 Kings xxiii. 12) which Josiah removed, were probably connected with the worship of the stars, and they prove the adoption of Babylonian usages. Amongst these we may imagine that "the dial of Ahaz" held a conspicuous place; but what its actual form was, must ever remain a matter of conjecture. The word "degrees" in our translation of the Scripture might have as appropriately been rendered "steps;" and it has generally been supposed that a pillar outside the king's palace threw a shadow on the steps of the terraced walk which indicated the time of day.

INTRODUCTORY.

It appears that in Egypt, the land of earliest civilization, obelisks and pillars were used for this end; but such contrivances would only show the progression of the natural day, from the rising to the setting of the sun. A yet more primitive mode of computing time is even now pursued in Egypt, and may have been in existence before the pillar-dial was ever erected. We are told that in Upper Egypt the natives plant a palmrod in the open ground, and arrange a circle of stones round it—forming a sort of clock face—and on this the shadow of the palm falls and marks the time of day. The plougher will leave his buffalo standing in the furrow to consult this rude horologe, and learn how soon he may cease from his work—illustrating the words of Job (vii. 2) "as a servant earnestly desireth the shadow."

The Rev. W. B. Galloway has proposed a construction of Ahaz's dial, quite different from that which has been already named. He thinks it may have been made on the plan of the huge Indian dials, such as were erected about 300 years ago at Benares, Delhi, and Agra, for the purpose of restoring the ancient sciences of Hindostan. These are assumed to be reproductions of the original pattern introduced from Babylon. Berosus, the Chaldean, who went to Athens, as some say, in the reign of Alexander the Great, was the great astronomer of his age; and he made a monster dial in form of a concave hemicycle, which Mr. Galloway thinks may have been the shape perpetuated in India. It was in fact a building of large size, and included a staircase leading to an observatory at the top, which formed a gigantic gnomon that cast its shadows on the coping of a wall below, which was built in the form of a hemisphere. But enough of guesses about the dial of Ahaz.

The introduction of dials into Greece is said by some to have taken place about the year 560 B.C. by Anaximander of Miletus, the successor of Thales. About 300 years later, Berosus seems to have taught the art of making dials of semicircular form, like the one that was found at the base of Cleopatra's Needle at Alexandria, and is now deposited in the British Museum. Vitruvius says, lib. ix. cap. 9, " Berosus, the Chaldæan, was the inventor of the semicircle, hollowed in a square, and inclined according to the climate." The old dials seem to have borne various forms. Some were suspended, and would require adjustment before they were consulted; and one of this kind, shaped like a ham, has been found at Herculaneum—the Romans having succeeded to a pattern, which certainly could not have had a *Jewish* origin. Lord Elgin brought a fixed dial from Athens. It is wrought in stone, and has four faces, each of which is lineated and numbered. It is supposed that it stood in one of the crossways of the city, and told the time to all comers in each direction. The octagonal Tower of the Winds at Athens contained a dial on every face.

As the Greek numerals are represented by the letters of the alphabet, it is curious that those letters which express the hours six, seven, eight, nine, (from noon till four o'clock) should spell the word ζηθι "live," that is, "enjoy thyself." An epigram, attributed to Lucian, comments upon this with the observation, " Six hours are enough for work; those which follow show by their very letters that we should then begin to enjoy ourselves." In fact, these hours form that portion of the day when exertion is scarcely possible in a warm climate, and people generally sleep and relax their bodies.

The Romans adopted dials from the Greeks, and the first erected at Rome was placed by Papirius Cursor in the court of the Temple of Quirinus, 293 B.C. Before this time "noon" was proclaimed by a crier, when the sun appeared between the rostrum and a spot called the "station of the Greeks." About thirty years afterwards, during the first Punic war, Valerius Messala captured a dial at Catania, in Sicily, which he sent to Rome, where it was placed on a pillar near the rostrum, and remained there for ninety-nine years, when Martius Philippus substituted another which told the time more accurately. The commoner form appears to have been merely that of a column which formed the gnomon, and threw its shadows on the ground; in fact, this was the most primitive mode of ascertaining the sun's rise and decline.

A learned friend offers the following remarks. "The shadow of a tree or vertical pillar cannot permanently indicate the time of day, because its motion is not uniform. The sun's motion in his diurnal track is uniform; he always describes the same angle in the same time; but the angular velocity of the shadow of a tree or pillar is greater at noon than it is at sunrise or sunset; it also varies with the time of year. The gnomon that indicates the time of day must slope to the horizontal plane at an angle equal to the latitude of the place, and must also lie due north and south. This may be illustrated by the blunder the Romans made in bringing a Sicilian sun-dial to Rome." (Pliny N. H. vii. 214, Censorin. de D. N. 23.) The same authority proceeds to say, "The proper slope of the gnomon may be obtained without a knowledge of the latitude; and the Babylonians probably did obtain this, and from it determined the latitude, and ascertained that the earth is spherical; so also the Greeks. (Strab. ii. pp. 125-136.) A vertical gnomon may be used to determine, not the time of day, but its length and variation of length in terms of equinoctial hours; and thus the Egyptian obelisk brought to Rome by Augustus was used. (Plin. N. H. xxxvi. 72.) Though from causes which Pliny conjectures, the inferences they drew were subsequently found to be erroneous. During the Attic period, the Greeks of that city ascertained the time of day by measuring a shadow; but it is difficult to determine how they did this. They talk of a six-foot shadow or mark, a ten-foot shadow or mark, &c. Expressions of this kind are very frequent, and yet they give little or nothing whereby to show the particulars of the measurement—whether it was the length of the shadow that was measured, or its angular distance from a given line, or even what the thing was that gave the shadow." [In Aristophanes is found the expression στοιχεῖον δεκάπουν, *a gnomon ten feet long*; and in other Greek writers of a later period the same word is used, with epithets signifying six, twelve, and seven feet. There also occurs the word ἡ σκιά, *the shadow*, to which the same epithets are applied.] "There is little in any of these writers to suggest even a conjecture, still less to support a probable one respecting the mode of measuring the shadow. The shadow was thrown on the ground; it was twenty feet long in the morning, about six at noon, and ten or twelve in the afternoon. Salmasius conjectures that it was each man's own shadow which he measured with his own foot. This is really ingenious; but all that is certain is, that the method was far from exact, very imperfect, and required altering several times in the year."

Such is the conclusion at which our learned friend has arrived; but one more quotation must be given from his kindly comments: "There certainly is a considerable probability that what is called poetic astronomy is as old as human nature itself; and it is a very perfect system. Without any instrumental aid the first occupiers of Arabia could determine the time of year and the time of day with as much accuracy as they had any occasion for. The loss of this science, and the causes, moral and historical, that produced it are curious, and as connected with the Holy Bible, they are important; but all these matters require leisure, long life, and patience—things which few possess, and still fewer wish for."

It is time that we descended from the heights of conjecture to the plain level of facts; remarking, by the way, that the studious contemplation of the heavenly bodies led to the worship of them, and also to astrology, which was a base corruption of the highest science known to men.

That dials were of frequent occurrence in ancient Rome is obvious from the lines attributed to Plautus, who died about 184 B.C.; and it is probable that their existence, or rather information, was noisily announced at stated intervals by trumpeter or crier.

> " The gods confound the man who first found out
> How to distinguish hours—confound him, too,
> Who in this place set up a sun-dial,
> To cut and hack my days so wretchedly
> Into small pieces! When I was a boy,
> My belly was my sun-dial—one more sure,
> Truer, and more exact than any of them.
> The Dial told me when 'twas proper time
> To go to dinner, when I had aught to eat;
> But, now-a-days, why even when I have
> I can't fall to, unless the sun gives leave.
> The town's so full of these confounded dials,
> The greatest part of its inhabitants,
> Shrunk up with hunger, creep along the street."

In the time of the Emperor Trajan, who died A. D. 117, the art of dialling must have been well understood, if an epigram, attributed to the Emperor, be authentic (See Anthol. Pal. xi. 418): ΤΡΑΙΑΝΟΥ ΒΑΣΙΛΕΩΣ. ’Αντίον ἠελίου στήσας ῥίνα καὶ στόμα χάσκον, δείξεις τὰς ὥρας πᾶσι παρερχομένοις— " Set your nose and wide mouth to the sun, and you will tell the hour to all passers by." He was ridiculing a man who had a long nose and a wide mouth, very much curved and grinning; whilst his many teeth, all visible, resembled the characters that denote the hours, and their double line. Prescott tells us that the Peruvian Indians had erected pillars of curious and costly workmanship, which served as dials, and from which they learned to determine the time of the equinox. When the shadows were scarcely visible under the noontide rays, they said that " God sat with all his light upon the column." Their Spanish conquerors threw down these columns, as savouring of idolatry.

Mahometan countries abound in these instruments, which are probably no new introductions. As prayer is ordered to be observed five times in every twenty-four hours, all the principal mosques in Constantinople are provided with a dial, in order that the people may ascertain the exact times of worship. The sun-dials on the mosques of S. Sophia, Muhammed, and Sulimania, have no motto or inscription, except what expresses the course of the shadow and the name of the maker. But on some, in addition to the lines which mark the solar movement, there is a line drawn which points to the sacred town of Mecca, towards which the faces of the faithful must be turned during the performance of their religious offices. It is said that the Turks erect a sun-dial, whenever they build a mosque.

"Sun-dials," writes a correspondent of the highest authority, "are the commonest things possible in China. You cannot get into your chair, or palanquin, but a flat board, with a dial fixed in the centre, is put before you to keep you in. They are on the sides of houses, and on boxes—indeed, they are most common, but none of us recollect any mottoes under them: though the Chinese have such a habit of putting mottoes to everything, that it is more than likely that sun-dials are no exception. They are probably ancient. There are sun-dials in Japan, for I had one in my garden." Touching Japanese dials, one who was for long resident in Japan, writes: "In regard to sun-dials, I can only say that there are sun-dials in Japan, but not as fixtures; and that they are not provided with mottoes, as is the case on old sun-dials in Europe. You will probably remember the small bronze portable sun-dials every Japanese carries about with him; but I never saw a large fixed sun-dial anywhere, except at a watchmaker's shop in Yokohama, who had made use of the railing round his shop as a kind of dial, according to which he adjusted his watches. The shadow of the railing had been previously adjusted, and was marked off after the Saturday gun from the flagship."

We may here remark that at Paris, and we believe also at Edinburgh and elsewhere, a cannon has been used for proclaiming the hour of noon, which was fired by the rays of the sun being concentrated on a magnifying glass so placed as to ignite the powder in the touchhole, when the sun reached its meridian height. Moreover, the gun stood on a platform which was marked as a sun-dial, and therefore simultaneously with the explosion, the gnomon cast its shadow exactly on the figure xii. We hardly need add that this mode of ascertaining 12 o'clock is not pursued at Greenwich or any scientific observatory; but that telescopic enquiry more accurately informs the Astronomer Royal when the sun attains its meridian.

No science being required for the construction of the common dial, we may expect to find barbarous nations measuring the progress of the day; and no doubt there are ancient pillars in various countries, the original purpose of which is unknown, but which were used as sun-dials. There are some very old pointed stones near Boroughbridge, in Yorkshire, still called the "Devil's Arrows," which may have been time-keepers; and certainly the early inhabitants of Great Britain could not have been behind their neighbours in horology, if it be true, as stated, that Julius Cæsar brought sun-dials and *clepsydræ* to this country. The Saxon dials at Kirkdale and Edstone in Yorkshire, and that at Bishopstone in Sussex, are the oldest remaining dials known in England, and all seem to belong to the eleventh century. Stone dials of a much earlier date have been found in Ireland.

Of all dials we have met with, none approach in architectural interest to those in Scotland, which appear to date from about 250 years ago. They will bear comparison for elegance and beauty with the wayside crosses of the middle ages. It is difficult to ascertain whence their style and form were derived. No constructions remaining in France can have suggested them. What is called "Queen Mary's Sun-dial" at Holyrood Palace, is fine; and was erected by Charles I. Another elegant dial is now, after removal, at Melville House, the seat of the Lady Elizabeth Cartwright, daughter of the Earl of Leven and Melville. Both these dials are on steps, and are finely carved; and it is in the hollows of the ball at the top of each, that the gnomon indicates the hours. The most remarkable, however, is the dial at Glamis Castle, the residence of the Earl of Strathmore, near Forfar, where, we are told, "the Pretender slept in 1715, and had above eighty beds made for himself and his retinue." Here was the Castle, by inheritance, of Macbeth—

> "*1st Witch*. All hail, Macbeth! hail to thee, thane of Glamis.
> *Macbeth*. Stay, you imperfect speakers, tell me more.
> By Sinel's death, I know, I am thane of Glamis."

This wonderful dial is supposed to have been made about the beginning of the seventeenth century. It stands on steps, and four carved lions above the base (Lyon is the family name) stand up, and hold each a shield in his paws which is a dial face. The names of months and days are engraved below. But as the structure tapers upwards, there are literally eighty dial faces cut diamond-wise on the blocks of stone, which look as if they had been carefully sliced over, to afford planes in which the gnomons are fixed.

In the reign of Elizabeth the mortuary crosses were cut down, or stumped, in our churchyards; and nothing is more common than to see these converted into dials: a brass plate and gnomon being placed on the shortened pillar. In the parish records of Prestbury there is the following entry: "1577. Item. for cuttynge the cross in the churchyard, and for charge of one with a certificate thereof to Manchester, xij^d."

In many instances, both of our cathedrals and parish churches, the clock has taken the place of the dial, which was generally fixed over the south porch on the wall. It was so at York Minster, for where the clock now shows its face over the south entrance, there was formerly a sun-dial, as exhibited in a plate in Dugdale's "Monasticon," now open before us.

The art of dialling, under the title of "Gnomonics," was taught in works of deep mathematical calculation, especially in the seventeenth century. A lively mathematician, recently opening a volume on this subject by William

Leyburn, published in 1682, exclaimed that "the five first books of Euclid were easy reading after that!"

We cannot presume to touch upon the intricate subject of horology by mechanism, but it is certain that clocks were invented long before they came into general use, and displaced the dial. More than 500 years ago, according to Dante, a clock that would strike the hours was known in Italy, and striking clocks were made in England at almost as early a date. A timepiece that could be carried on the person was also a common implement in Shakespeare's time; but we cannot suppose this to have been a watch with metal works. Jaques tells us that his motley Fool

> "Drew a dial from his poke:
> And looking on it with lack-lustre eye,
> Says, very wisely, *It is ten o'clock*."

The peasant in the Pyrenees can do the same thing now, with possibly a like time-teller. He carries in his pocket a small cylinder, made of boxwood, and not larger in size than a pocket-knife. The top of it can be drawn out, when a small blade turning on a pin forms a gnomon, which can be adjusted to the lines, figures, and initials of the month that are carved in the wood. It will tell the time when consulted, within five minutes. We suggest this form as more simple and primitive than the ring-dial, which some think was the article alluded to in "As you Like it." The ring-dial was much used in the seventeenth and at the beginning of the eighteenth century, and its inside was marked like a dial-face. A hole in the ring admitted the ray of light, which passing through fell on the interior surface; and there was a slide which required adjustment for ascertaining the hour.

There have been many quaint devices connected with dials. For instance, in the garden of Wentworth Castle, near Barnsley, the property of F. Vernon Wentworth, Esq., a dial was formed of box edgings cut into the proper numerals, whilst a clipped yew tree in the centre acted as the gnomon. Floral dials have also been invented, being composed of flowers that bloomed in succession during the months of sunshine. These, however, are conceits, which hardly come within the compass of our subject.

As clocks ascended into the church towers, or showed their faces in the market-places, the dialler's learned vocation gradually ceased. The old dial may still retain its footing in the quaint yew-tree'd garden, or may stand conspicuously in the churchyard; but few consult it as an oracle, and it rather lingers superfluously amongst us as a memento of the past. It has nevertheless to many minds a touching interest; it has drawn forth maxims in the form of mottoes, and it would be like discarding wisdom if we did not preserve and cherish them.

> "But if these shadows tell us after all
> We are but shadows on Life's sunny wall:
> They not less point us, with a hope as bright,
> To that good land above where all is light."
>
> *H. V. T.*

Howard, the philanthropist, is said to have thus spoken on his deathbed:—"There is a spot near the village of Dauphigny where I should like to be buried. Suffer no pomp to be used at my funeral; no monument to mark the spot where I am laid, but put me quietly in the earth, place a sun-dial over my grave, and let me be forgotten." Sir William Temple ordered that his heart should be placed in a silver case, and deposited under the sun-dial in his garden at Moor Park. So tender have been the uses to which the dial has been applied, so striking is the thought that the eyes of succeeding generations look in its time-telling face only to read their own *memento mori*, that we are ready to fall into David Copperfield's vein of meditation, as we see it ever cheerfully return with sunlight to the performance of its duties, and ask, "Is the sun-dial glad?—I wonder, that it can tell the time again."

* * * * * *

INTRODUCTORY.

We can imagine a strict diallist, after reading the foregoing sketch of the history of dials, exclaiming, "You do not say what a dial is; and as not one in a thousand knows this, explanation of it might be fairly expected."

From his point of view, which would include the lineating and fixing a dial in the proper manner and right position, the accusation would be just; but as we have altogether declined to touch the scientific side of the subject, and have only undertaken to treat what is obvious to the senses, we cannot be censured for not venturing to approach the astronomical and mathematical mysteries. Holding by this rule, we maintain that it cannot surely be true that the world in general, more especially the literary portion of it, for whom we write, can be so ignorant of what their forefathers knew, that they have to be told what those primitive horologes—sun-dials—are! Nevertheless, we will offer a slight description. A sun-dial is a timepiece of shadows, and in its first idea was so simple, that as Charles Lamb said, "Adam could scarce have missed it in Paradise." But those were days of rough reckonings, when there was little to be gained or lost by being a few minutes early or late. What then is a sun-dial now? Well, it is still a timepiece of shadows; but instead of the shadows being thrown from trees, pillars, or buildings requiring a large extent of space, we have, as it were, gathered them up into the small compass of a foot or two of level board, producing them by a bar of iron or wood raised at a proper angle from the surface. These dial plates are marked round by regular lines of division, which show the places in which the shadow will fall at each successive hour; and, indeed, agreeably to the need of the times, the sixty minutes of each hour were soon marked off also.

How this was accomplished, how the calculations were made which enabled the diallist to set up a dial at any aspect with equal certainty of telling the hours correctly, we do not pretend to describe. This part of the subject belongs to scientific enquiry, and we must refer the readers, curious on this point, to such old books as "Leadbetter's Art of Mechanic Dialling," &c., for information.

The foregoing account, brief as it is, will, it is hoped, be enough to make a fair beginning by explaining what a sun-dial is—perhaps more than enough, for we cannot believe that the numberless clergy with sun-dials on their church porches have never once looked up at them, and found out what they were. Nor that clergymen's wives have marshalled their numerous families into church, Sunday after Sunday, for so many years, and no little fingers pointed to, and no little tongues asked about the pretty tablets, often shining with the colour of the gilding over their heads, as they entered the sacred building.

Besides, England is not the world, nor is its climate favourable to horologes dependent on sunshine. Go to Nice, Antibes, Cannes, and probably other towns and villages in the south of France, and you will find them on all sorts of buildings, prettily coloured, fancifully imagined, on two or three houses in the same street, on the toll-house of a bridge here, on a shed there. They are common along the Riviera from Genoa to Nice, as is evident from the collection of mottoes inserted by the late Dean Alford in his beautiful book, "The Riviera." Moreover, did not H. R. H. the Prince of Wales visit the Temple in 1861, and on this occasion were not all the old Temple dials done up, that is, regilt and repainted, to greet his presence?

No, the proportion of those ignorant of what a sun-dial is, cannot even in the present day, when they are jostled out of use by chronometers, be as one to ninety-nine.

The present collection of dials, with their mottoes, was begun long ago —according to the measure of a man's life—some thirty-five or forty years about. Perhaps the presence of a curious old dial over our church porch (Catterick), with something like a punning motto, "Fugit hora, ora," may

INTRODUCTORY.

have had somewhat to do with starting the idea. Also at the home of some dear friends, a few miles off, the porch of their picturesque little church (Wycliffe), on the banks of the Tees, bore another inscription, "Man fleeth as a shadow." A third motto surmounted an archway in a stable-yard (Kiplin) "Mors de die accelerat." A fourth was over the door of a cottage in a village (Brompton-on-Swale), bearing the warning words, "Vestigia nulla retrorsum," which shone out in gold and colour amidst evergreens. Here lived the venerable sister of a canon of Lincoln, which may perhaps account for the presence of the dial. A fifth looked out from the depths of pyracanthus on a house at Middleton-Tyas, hinting to callers not to waste the precious hour, with its "Maneo nemini;" while last, and not least in our esteem, stood the touching inscription, "Eheu, fugaces!" on a pillar-dial outside the drawing-room at Sedbury Hall, Yorkshire, where it betokened the scholarly character of the hospitable owner. These six mottoes (all, rather remarkably, in one neighbourhood), made an admirable beginning of a list, which soon swelled to twenty or thirty pages by taking a wider circuit, and with the assistance of the contributions of friends. And thus the matter went on from more to more; but the great impulse was given when the friend, alluded to in the preface, undertook to collect in the south of France and the north of Italy—a fair field indeed, and one even yet imperfectly explored. As to these dial mottoes there are perhaps as many differences of opinion, as there are differences of character in those who read them. We, who have studied them for so many years, feel with Charles Lamb that they are often "more touching than tombstones," while to other people they seem flat, stale, and unprofitable. One correspondent describes them as "a compendium of all the lazy, hazy, sunshiny thoughts of men past, present, and in posse;" and says, "the burden of all their songs is a play upon sunshine and shadow." But this is no fair description. The poet's words,

> "Liberal applications lie
> In art as nature,"

have never been more fully realized than in the teachings which have arisen from dials, as we trust the following pages will prove beyond a doubt. So far from the burden of all their songs being a play upon sunshine and shadow, one of the most fertile subjects of thought is the sun's power, as being his own time-keeper, which he certainly is, whilst the mottoes constantly assert the fact.

The sun describes his own progress on the dial-plate as clearly as he paints pictures on the photographer's glass—human art assisting in both cases. "Solis et artis opus," says the dial in a street at Grasse, near Cannes —somewhat baldly perhaps. More refined is the "Non sine lumine" of Leadenhall Street; and perhaps higher still the "Non nisi cœlesti radio" of Haydon Bridge. "Non rego, nisi regar" is the modest avowal of another dial in a street at Uppingham, acknowledging itself to be but an instrument governed by an overruling power. And these are but a few of the many "applications" the poet speaks of. The reader will find all these mottoes in their proper places in the list, on which our first happens to be a foreigner.

SUN-DIALS.

CHARLES LAMB ON THE TEMPLE SUNDIALS.

"What an antique air had the now almost effaced sun-dials, with their moral inscriptions, seeming coevals with that Time which they measured, and to take their revelations of its flight immediately from heaven, holding correspondence with the fountain of light! How would the dark line steal imperceptibly on, watched by the eye of childhood, eager to detect its movement, never catched, nice as an evanescent cloud, or the first arrests of sleep!

>   Ah! yet doth beauty like a dial-hand
>   Steal from his figure, and no pace perceived!

What a dead thing is a clock, with its ponderous embowelments of lead and brass, its pert or solemn dulness of communication, compared with the simple altar-like structure and silent heart-language of the old dial! It stood as the garden god of Christian gardens. Why has it almost everywhere vanished? If its business use be superseded by more elaborate inventions, its moral uses, its beauty, might have pleaded for its continuance. It spoke of moderate labours, of pleasures not protracted after sunset, of temperance, and good hours. It was the primitive clock, the horologe of the first world. Adam could scarce have missed it in Paradise. It was the measure appropriate for sweet plants and flowers to spring by; for the birds to apportion their silver warblings by; for flocks to pasture and be led to fold by. The shepherd "carved it out quaintly in the sun," and, turning philosopher by the very occupation, provided it with mottoes more touching than tombstones."—*Elia*.

SUN-DIALS.

1.

A LUMINE MOTUS.

Moved by the light.

N a dial at Sestri Ponente, on the Riviera, a few miles to the west of Genoa. This, the first motto on our list, suggests some fanciful thoughts. Light moving a shadow—one intangible thing acting on another, and the result becoming visible. It can be explained, of course, in the most matter-of-fact manner; but this need not hinder our casting a fleeting glance at the idea. A curious list might be made of things most familiar to us, such as wind, thunder, &c., with the view of ascertaining by how few or how many of our senses they are cognizable.

2.

A ME TOCCA POI LA SORTE
DI SEGUIRTI FINO A MORTE.

My fate is to follow you to death.

At Graglia in Piedmont.

3.

A SOLIS ORTU USQUE AD OCCASUM.

From sunrise to sunset.

On the cemetery wall at St. Gervais in Savoy.

4.

A SPAN IS ALL THAT WE CAN BOAST,
AN INCH OR SO OF TIME:
MAN IS BUT VANITY AND DUST
IN ALL HIS FLOWER AND PRIME.

On an erect stone dial, south declining east, which has been recently placed over the door of a barn belonging to a large farm near Bedale, in Yorkshire, known as East Lodge. The initials of the owner, G. J. Serjeantson, Esq., with the date 1862, are below the dial, which was erected when the barn was built. Hear Wordsworth on the agricultural mind:—

> "The shepherd lad, that in the sunshine carves,
> On the green turf, a dial—to divide
> The silent hours; and who to that report
> Can portion out his pleasures, and adapt,
> Throughout a long and lonely summer's day,
> His round of pastoral duties, is not left
> With less intelligence for *moral* things
> Of gravest import. Early he perceives
> Within himself a measure and a rule,
> Which to the sun of truth he can apply,
> That shines for him, and shines for all mankind."
>
> *Excursion*, B. iv.

5.

AB HOC MOMENTO PENDET ÆTERNITAS.

On this moment Eternity depends.

This favourite motto may be seen at Sprawley, in Worcestershire; also on a wooden dial fixed against a house at Offerton, between Stockport and Marple; and in a private garden at Northampton.

6.
AB ULTIMA CAVE. 1838.
Guard against the last hour.

Painted, with the date, 1838, on a dial over the door of a house which stands surrounded by trees near the edge of a cliff facing the rocky headland of Porto Fino, in the Gulf of Genoa. There is a local tradition that an English queen was buried here. Berengaria, queen of Richard I., suffered shipwreck in the Mediterranean; which was perhaps enough to give rise to a legend for which there seems to be no historical evidence.

7.

ABUSE ME NOT, I DO NO ILL:
I STAND TO SERVE THEE WITH GOOD WILL;
AS CAREFUL THEN BE SURE THOU BE
TO SERVE THY GOD, AS I SERVE THEE.

This inscription was on a copper dial in Crompton churchyard, in the parish of Oldham. The mortuary cross on which it was erected still remains, but the dial-plate was stolen a short time ago. It is, however, intended to put up a new dial with the old motto upon it.

8.
ÆTAS CITO PEDE PRÆTERIT, 1787.
The age passes with a swift foot.

This is written on a dial on the porch of the Church of St. Hilary, near Marazion, in Cornwall, which has been recently and beautifully restored by Mr. W. White, F.S.A. The motto is peculiarly interesting from the fact that the venerable and respected vicar, the Rev. Thomas Pascoe, was instituted to the benefice so long ago as in 1814. *Marazion*, or the Bitter-Zion, is the name given to the place by some Jews who settled there, and formerly held markets in the town for the sale of tin,—hence, by the country-folk, it is called *Market Jew*. In old times it was much supported by pilgrims visiting the shrine of S. Michael.

9.

AH, WHAT IS HUMAN LIFE!
HOW LIKE THE DIAL'S TARDY MOVING SHADE:
DAY AFTER DAY GLIDES BY US UNPERCEIVED,
YET SOON MAN'S LIFE IS UP, AND WE ARE GONE.

On a dial at Hesketh, in Lancashire.

10.

AFFLICTIS LENTÆ, CELERES GAUDENTIBUS HORÆ.

The hours are slow to the afflicted, swift to the joyous.

On a dial at Courmayeur, in the valley of Aosta; also at Hyères. Dean Alford met with the same between Bordighera and the Nervia, and thus renders the meaning:—

> "To them that mourn the hours are slow,
> But with the joyful swiftly go."

This latter sentiment is corroborated by W. H. Spencer, the poet of the boudoir:—

> "Too late I staid, forgive the crime,
> Unheeded flew the hours:
> For noiseless falls the foot of time
> Which only treads on flowers."

11.

ALLEZ VOUS.

Pass on.

Some years ago, a Dutch vessel came into port at Dartmouth, and brought a Dutch sun-dial of singular workmanship which bore this motto. The dial came into the possession of the Vicar of St. Petrox, Dartmouth, and was placed at the time in the garden of the vicarage; but it is no longer there. The sentiment of the inscription seems to be a reproof to idle lookers on, bidding them begone about their proper business. It is curious to compare this with motto No. 22; which may, perhaps, have been a real dial motto after all.

12.

AMBULANTES IN NOCTE AUDIEBANT VOCEM DEI.

Walking in the night they heard the voice of God.

Was on a sun-dial in the garden of Dr. Young, author of "Night-Thoughts," at Welwyn, in Hertfordshire. The motto, though impressive, is somewhat irrelevant, but characteristic of the inscriber.

13.

AMICIS QUÆLIBET HORA.

Any hour you like, for friends.

At Grasse, in the Department of the Alpes Maritimes, on a house which is situated in one of the small old-fashioned squares of the town. It is also inscribed on a dial that is sketched out on the yellow wall of a house in Murano, which stands in a garden beside one of the canals that intersect the island. The shape of the dial, which is dated 1862, is oval; and the hour xii. is marked by the figure of a bell.

In addition to the Latin inscription, there is the following Italian motto written below:—

>   L'ORA CHE L'OMBRA MIA FEDEL TI ADDITA,
>   PENSA CHE FU SOTTRATTA ALLA TUA VITA.

The hour which my faithful shadow points out to thee,
Remember, has been taken from thy life.

The dial is placed between the upper windows, and just escapes the shadow cast by the low roof; but occasionally it is eclipsed, even on the brightest day, when the west wind blows the smoke across it from the famous glass-works of Murano.

14.

ARRIPE HORAM, ULTIMAMQUE TIMEAS. 8BRE 1812.

Snatch the (present) hour, and fear the last.

Is on a house at Tours. The dial is more fanciful than useful. It is square in shape, and at the two upper corners the sun and moon are figured. On the dial-face there is no numeral but xii., which is at the bottom, and the gnomon is formed by two arms projecting from a crescent: they are centrally placed, and lay hold of the sun.

15.

AS A SHADOW, SUCH IS LIFE.

Is on a dial placed immediately over the entrance of the south porch of Wensley Church in Wensley Dale, Yorkshire.

16.

AS TIME AND HOURS PASS AWAY,
SO DOTH THE LIFE OF MAN DECAY:
AS TIME CAN BE REDEEMED WITH NO COST,
BESTOW IT WELL, AND LET NO HOUR BE LOST.

These lines are engraved on a very elaborate piece of workmanship, which forms both a dial and compass, and is preserved in the Museum at Edinburgh. It bears the form of a hunting watch when opened; and on one side the meridians " of all the principall townes and cities of Europe " seem to be inscribed; and there is also written, " This table beginneth at 1572, and so on for ever."

17.

ASPICE UT ASPICIAS.

See that you see.

Is on a dial which rests on a four-sided pillar in the churchyard of Areley Kings, Worcestershire. On the sides of the pillar are other inscriptions, with sculptures. The south side has a figure of Time, and is inscribed:—

> TIME'S GLASS AND SCYTHE
> THY LIFE AND DEATH DECLARE,   CONSIDER
> SPEND WELL THY TIME, AND
> FOR THY END PREPARE.

On the north side there is a figure of Death, with darts, hour-glass, and shovel; and these lines are given:—

> THREE THINGS THERE BE IN VERY DEEDE,
> WHICH MAKE MY HEART IN GRIEF TO BLEEDE:
> THE FIRST DOTH VEX MY VERY HEART,
> IN THAT FROM HENCE I MUST DEPARTE;
> THE SECOND GRIEVES ME NOW AND THEN,
> THAT I MUST DIE BUT KNOW NOT WHEN;
> THE THIRD WITH TEARS BEDEWS MY FACE,
> THAT I MUST DIE NOR KNOW THE PLACE.
>
> J. W. FECIT ANNO DṀNI, 1687.
>
> BEHOLD MY WILLING DART AND DELVING SPADE;
> PREPARE FOR DEATH BEFORE THY GRAVE BE MADE;
> FOR
> AFTER DEATH THERE'S NO HOPE.
>
> IF A MAN DIE HE SHALL LIVE AGAIN.
>
> ALL THE DAYS OF MY APPOINTED TIME
> WILL I WAIT TILL MY CHANGE COME. Job xiv. 14.
>
> THE DEATH OF SAINTS IS PRECIOUS,
> AND MISERABLE IS THE DEATH OF SINNERS.

On the east side of the pillar is the following:—

    SI VIS INGREDI IN VITAM,
    SERVA MANDATA.

If thou wouldst enter into life, keep the commandments.

    JUDGMENTS ARE PREPARED FOR SINNERS. Prov. xix. 29.

On the west side there is:

    SOL NON OCCIDAT
    SUPER IRACUNDIAM VESTRAM.

Let not the sun go down upon your wrath.

    WHATSOEVER YE WOULD THAT MEN
    SHOULD DO UNTO YOU,
    DO YE EVEN SO UNTO THEM.

.18.

ASPICIENDO SENESCIS.

Whilst looking you grow old.

This motto is painted on the south wall of a house in the Place Bellevue, Nice, near the old port. It is close under the roof, much wider than long in shape, and has a red border. The words are underneath the dial in capitals. Part of the plaster having fallen off, the date has been left imperfect: "26 Giugno 182—." The same inscription is on an upright dial on a house wall in one of the squares at Aix-les-Bains, Savoy, dated 1853.

19.

AUJOURD'HUI A MOI,
DEMAIN A TOI.

To-day for me, to-morrow for thee.

In the cemetery of Courmayeur.

20.

AUT DISCE, AUT DISCEDE: MANET SORS TERTIA, CÆDI.

Either learn, or go away: a third choice remains, to be flogged.

The first line of the above is a dial-motto at the Royal Military College, Woolwich; and we add the fuller inscription which is preserved on an old tablet at the end of the schoolroom of Winchester College, being characteristic of the hardy discipline of that noble public school.

21

AUT LAUDA, VEL EMENDA 1738, R. Nellson, Fecit.

Either commend, or amend.

The note respecting this dial has been imperfectly filled up; and the collector has lost all recollection of the locality. Its application is dubious.

22

BEGONE ABOUT YOUR BUSINESS.

Is inscribed on a wooden dial of a house at High Lane, near Disley, in Cheshire. Mr. Timbs records that it was on the dial of the old brick house which stood at the east end of the Inner Temple terrace, and was removed in 1828. The brusqueness of the advice is accounted for by the following pleasant legend given in " Notes and Queries," 2nd. S. v. ix. p. 279. "When the dial was put up, the artist enquired whether he should (as was customary) paint a motto under it? The Benchers assented, and appointed him to call at the Library on a certain day and hour, at which time they would have agreed upon the motto. It appears, however, that they had totally forgotten this; and when the artist or his messenger called at the Library at the time appointed, he found no one but a cross-looking old gentleman poring over some musty book. 'Please, sir, I am come for the motto for the sun-dial.'

'What do you want?' was the pettish answer: 'why do you disturb me?' 'Please, sir, the gentleman told me I was to call at this hour for a motto for the sun-dial.' '*Begone about your business!*' was the testy reply. The man, either by design or mistake, chose to take this as an answer to his enquiry, and accordingly painted in large letters under the dial, BEGONE ABOUT YOUR BUSINESS. The Benchers when they saw it decided that it was very appropriate, and that they would let it stand—chance having done their work for them as well as they could have done it for themselves. Anything which reminds us of the lapse of time should remind us also of the right employment of time in doing whatever business is required to be done." The idea is repeated on the church at Kilnwick-on-the-Wolds, Yorkshire.

23.
BREVIS HOMINUM VITA.
Man's Life is short.

The dial-motto on the church at Aberford, in Yorkshire.

24.
BULLA EST VITA HUMANA.
Human life is a bubble.

One of several mottos on a Cross-dial at Elleslie, near Chichester.

25.
CADE L'OMBRA AI RAI
NEL MEZZO GIORNO,
E SINO ALL' OCCASO
IL LOR SOGGIORNO. 1853.

The shadow falls under the rays (of the sun) at noontide, and until sunset is their sojourn.

Alluding to the position of the dial which faces west, and therefore catches the sun's rays from midday to sunset. It is on a house at La Tour, the little capital of the Vaudois valleys of Piedmont, and is very plain in design. The motto is in the corner, and the date below. The dial is painted on the wall, high up betwixt the windows, which have balconies, and look into a small square. In front are the mountains which witnessed so many tragedies during the Vaudois persecutions. Close to this house is the former residence of Colonel Beckwith. This officer, who served under the Duke of Wellington in the Peninsular war, from accidentally reading Dr. Gilly's "Waldensian Researches," whilst waiting for an interview with the Duke at Apsley House, became so deeply interested in the Vaudois Church and people, that he resolved to dedicate the remainder of his life and fortune to their cause. He forthwith settled at La Tour, and having, what he called, "a brick and mortar gift," commenced building schools, colleges, hospitals and churches throughout the district. No less than one hundred and twenty schools were eventually erected, mainly through his instrumentality, in different parts of these valleys. "Let the name of Colonel Beckwith be blessed by all who pass this way," is inscribed over one of his schools.

.26.

CARPE DIEM. 1855.

Snatch the (passing) day.

This advice of Horace is on a gable of one of the wings of Heslington Hall, near York, the seat of G. J. Yarburgh, Esq. The house is an old Elizabethan mansion, and was restored in 1855. The dial is circular in form, and painted blue and gold. The same motto occurs on a dial in High Street, Lewes, close to the County Hall, and is dated MDCCXVII. It is also painted on the tower of Offchurch, Warwickshire; where there is a full-faced sun gilt, with the gnomon as a nose: dated 1795, "William Snow, Churchwarden."

27.

C'EST L' HEURE DE BIEN FAIRE.

This is the hour for doing good.

This dial motto appears at Nice, and Dean Alford also found it at Porto Maurizio.

28.

CETTE OMBRE SOLAIRE EST A LA FOIS LA MESURE DU TEMPS, ET L'IMAGE DE LA VIE.

This solar shadow is at once the measure of time, and the symbol of life.

At Courmayeur.

29.

CHEMINEZ TANDIS QUE VOUS AVEZ LA LUMIERE. 1668.

Walk while ye have the light.

UT HORA SIC VITA.

As an hour so is life.

היום קצר והמלאכה מרובה

The day is short and the work great.

These three inscriptions are on a dial on the church wall at Hatford, near Farringdon, Berks. The Hebrew one (which is ill cut and partly defaced) is from the Talmud.

30.

CITO PEDE PRÆTERIT ÆTAS. 1679.

The age goes by with speedy foot.

On a dial in Over Peover churchyard, Cheshire; also in a garden at Woodhouse Eaves, near Loughborough.

31.

CITO PRÆTERIT ÆTAS.

The age passes swiftly.

On Lincoln Cathedral, which has also the common motto, *pereunt et imputantur*. The inscriptions are on separate dials, which are fixed on the two sides of the south-east buttress of the east transept of the cathedral, and are probably about two hundred years old.

32.

CONCITO GRADU.

With rapid step.

To be read on the tower of Rushton Church, Staffordshire, apparently the inscription of a worn-out dial.

33.

CORRIGE PRÆTERITUM,
PRÆSENS REGE,
CERNE FUTURUM.

Correct the past, direct the present, discern the future.

In the Altmarkt at Dresden.

34.

CORRO A QUEL DI CHE DEL SIGNOR LA SPADA
UNA FARA L' ITALICA CONTRADA:
QUESTA FIDA CHE FA LANCE CH' IO PORTO
SEGNA L' ORA D' UN POPOLO RISORTO.

I run to that day when the sword of the Lord shall unite the Italian country. This faithful [sword] that I bear, which forms the gnomon, marks the hour of a resuscitated people.

This motto is political, and a literal translation of it is most difficult. It has puzzled not only good Italian scholars, but native Italians also. The above rendering is a conjecture. There is some obscurity about the word *lance*. The dictionaries give *lancetta* as used for the gnomon on a dial; possibly, therefore, the gnomon in this case bore the form of a sword, and *questa fida* may be referred to *spada* in the first line. The inscription is placed above two dials, which stand side by side on the cathedral wall of Chieri, in Piedmont. One of them also shows the meridians of the chief cities of the world. Chieri is a few miles from Turin, on the left bank of the Po. It is an old town, but has suffered too much in the mediæval wars to retain many vestiges of antiquity. It has a round church of early Lombard architecture, which is now used as a baptistery. In its brighter days it was a free town, sending traders over half of Europe. It often changed its protectors: sometimes from choice, but more frequently from necessity; and at last gave its allegiance to the Counts of Savoy in 1347. The family of Balbo springs from Chieri; and one branch of this house, the Bertoni, refusing to accede to the treaty of 1347, emigrated to Avignon, where they assumed the name of Crillon, and were ancestors of the "Brave Crillon," *l'homme sans peur*, or *le brave des braves*, as he was called when serving under Henry IV. The other branch remained Piedmontese, to the glory and benefit of their country.

35.

COSI LA VITA.

Such is life.

To be read at Albizzola.

36.

CUM RECTE VIVAS, NE CURES VERBA MALORUM.

When you are living aright, you do not care what the wicked say.

This dial motto is to be seen at Poirino, in Piedmont.

37.

CUM UMBRA NIHIL,
SINE UMBRA NIHIL.

With the shade nothing: without the shade nothing.

Is on a dial on the Italian douane of the Splugen Pass, near Campo Dolcino; modern, and painted on the wall in red and yellow. The motto is also at Castasegna, a frontier village on the road to the Engadine; and at Bececa in the Trentino, the scene of Garibaldi's battle in 1866, where by some curious misspelling, or possibly an old form, the first *nihil* is written *nichil*.

38.

DAMMI IL SOLE, E DEL GIORNO L' ORA E CERTA;
SOLO DEL UOMO E L' ULTIMA ORA INCERTA.

Give me the sun, and the hour of the day is certain; of man alone is the last hour uncertain.

Inscribed on the church wall at Arola, without date, and the dial is probably modern. Arola is a small village on a hill-side, shaded by large chestnut trees. It is situated between the Lago d' Orta and the Val Sesia; and the mule path from Pella to Varallo, across the Col di Colma, runs through its narrow roughly-paved streets.

39.

DAYS AND AGES ARE BUT AS A SHADOW OF THE
   ETERNAL;
BUT THEIR USE, O MAN, DETERMINES THY FUTURE
   WEAL OR WOE.

PULVIS ET UMBRA SUMUS.
We are dust and shade.
ΕΛΕΥΣΟΝΤΑΙ ΓΑΡ ΗΜΕΡΑΙ Κ. Τ. Λ.
For the days shall come, &c.

These mottoes are on the keep of Carlisle Castle, just above the magazine.

40.

DEATHE JUDGMENT HEAVEN HELL UPON THIS
MOMENT DEPENS ETERNITIE. O ETERNITIE
O ETERNITIE O ETERNITIE. 1658.

The foregoing is inscribed in several lines on what is called " Sir Francis Howard's Dial," at Corby Castle, near Carlisle. It stands on the lawn, before the house, carved in stone, and has four sides. Sir Francis Howard was the second son of Lord William Howard (a son of Thomas, fourth Duke of Norfolk), who was the "belted Will Howard" of Walter Scott's "Lay of the Last Minstrel:"

> "Belted Will Howard is marching here,
> And hot Lord Dacre with many a spear."
> *Canto* iv. 6.

Sir Francis Howard was born August 29, 1588, and died in May, 1660. He first married Margaret, daughter of John Preston, Esq., of the Manor of Furness, Lancashire; and secondly, Mary, daughter of Sir Henry Widdrington, of Widdrington Castle, Northumberland. On the so-called dial are four shields: one on each of the four sides. One side contains the family initials; another shows the emblems of the Passion in relief, namely, St. Peter's cock, the scourge, the crown of thorns, the cross and five wounds, with the hands, feet, and heart represented; the seamless garment also, and below it the dice; the manacles in the form of I H S in letters of the period, and the hammer and nails. On another shield are the arms of Howard and Widdrington impaled, and on the fourth is the motto.

41.

DEFICIT SOL, NEMO RESPICIT.

When the sun fails no one regards.

DISEGNA LE ORE SENZA FAR ROMORE.

It marks the hours without making a noise.

These two mottoes are inscribed on the wall of the Italian douane at Fornasette, between Lugano, and Luino on the Lago Maggiore. They are on a large erect south dial, which is elaborately painted with curves and flourishes. The form is that of a shield, and the letter M stands for the numeral xii., with the date below, "15 Maggio, 1839." It is placed near the corner of the house, and beneath it are openings into a sort of corridor, where the work of the Custom House is carried on.

42.

DELLA VITA IL CAMMIN L' ASTRO MAGGIORE
SEGNA VELOCE AL GIUSTO E AL PECCATORE.

The greater star (the sun) swiftly points the path of life to the just man and to the sinner.

A dial motto on the outside wall, facing the road, of the Convent della Quiete, near Florence. The convent is now used as a girls' school.

43.

DELLE LINEE MATEMATICHE SAPER L' ORA POTRAI,
SE DENSA NUBE NON COPRE DEL SOLE I RAI.

By the mathematical lines thou canst know the hour, if a dense cloud hides not the rays of the sun.

On a house at Caprile, Venetia, and also in other places in the Italian Tyrol.

44.

DER TOD IST GWISS, UNGWISS DER TAG,
VIELLEICHT DASS DIESE STUNDE SEIN MAG;
DARUM THU' RECHT, UND DUENKT DABEI,
DASS JEDE STUNDE LETZTE SEI.

*Death is sure, the day uncertain: perchance this is the very hour thereof;
Therefore do right, and think on this, that each hour may be the last.*

The sense of the second line, as it stands, is doubtful; but the probable meaning is expressed above. The dial is on Herr Weber's house at Schwyz. Some of the words are in old German, ill-spelt and imperfect. The design of the painting represents the Virgin with the Child in her lap. She is richly robed and crowned, and her head is encircled with stars.

45.

DETEGO TEGENDO.

By covering I reveal.

On the south wall of a house in the Rue d'Antibes, Cannes. The plaster on which the dial is painted having been broken off, the first syllable is obliterated. It was copied in 1860.

46.

DIALL (loq.).
STAIE, PASSINGER
TELL ME MY NAME
THY NATURE.

PASS. (resp.)
THY NAME IS DIE
ALL. I A MORTALL
CREATURE.

DIALL (loq.).
SINCE MY NAME
AND THY NATURE
SOE AGREE,
THINKE ON THY SELFE
WHEN THOV LOOKS
VPON ME.

There is an ancient dial, having four sides, at Millrigg, in the parish of Culgaith, near Penrith. The opening dialogue betwixt Dial and Passenger is inscribed on one side of the square, and on the other side is Dial's moral deduction from it. The two remaining sides of the square are occupied by the armorial bearings of the Dalston family with the initials "I. D.," and those of the Fallowfields with "H. F."

John Dalston resided and died at Millrigg in 1692. He was the son of Sir Christopher Dalston, of Acorn Bank, who was knighted by James I. in 1615. This latter place was the chief residence of the family. The manor of Temple Sowerby, immediately adjoining, was granted by Henry VIII to Thomas Dalston, Esq., on the dissolution of religious houses. It belonged originally to the Knights Templars, and afterwards to the Hospitallers. Millrigg is now occupied as a farm-house.

47.

DEUS ADEST LABORANTIBUS. 1742.

God is present to those who labour.

At Hermit Hill, Wortley, near Sheffield.

48.

DIDST THOU NOT SEE THE LORD, HOW HE EXTENDED THY SHADOW.

Is the translation of a verse of the Koran, which is inscribed on a dial erected by the astronomer, Ali Kushaji, near the mosque of Muhammed II., by the gate of the Dyers at Constantinople.

49.

DIE AUGEN DES HERRN SIND HELLER ALS DIE SONNENSTRAHLEN. 1480 (or thereabouts).

The eyes of the Lord are brighter than sunbeams.

On a church at Hallstadt; the dial being roughly painted on the wall.

50.

DYE JETZIGE STUND UND DAS ZYTLICHE GLUECK SCHLEICHT HIN IN EINEM AUGENBLICK. 1792.

The present hour and temporal happiness steal away in a moment.

The dial is on a house in the market-place at Thun, in Switzerland, placed over one of the low wide arches of the arcades, where the Bernese peasants congregate on market days. It declines S. W., and is unusually large. It is painted pink on the whitewashed wall, and is rendered somewhat indistinct by the weather. The hours are in Roman characters, and are written on a scroll. It is a venerable-looking dial, worthy of the picturesque little town.

51.

DIES AFFERT MULTA.

The day brings with it many things.

This inscription was cut on a dial, the work of an ingenious and well educated man for his time, named Daniel Rose, who placed it over the doorway of his cottage house, called "Shutts," near Ashopton, in Derbyshire. He was the clerk of Derwent Chapel, also a schoolmaster and dial-maker—possibly, too, he wrote letters for his neighbours, and made their wills. It is said that he carved his dials in a soft slate stone during schooltime with a penknife: the dials both at Derwent Church and the Hall are specimens of his skill. Father and son were the parish clerks in succession. The father of Daniel Rose was a Welshman, and his mother is reported to have attained the age of 105. The family is not extinct in the parish of Derwent.

52.

DIES DEUM DOCET, DISCE.

The day teaches a God, learn.

A four-faced dial over the porch of the old church at Barmston, near Bridlington, Yorkshire, is thus inscribed. The letters are partly illegible.

Many of them are so much defaced that it needed all the skill and ingenuity of a clever scholar to even conjecture what they must have been. The above is the result; but the late Rev. J. Eastwood, who came to this conclusion, did not speak with certainty.

53.

DIES MEI SICUT UMBRA DECLINAVERUNT.

My days have gone down like a shadow.

Is on the Roman Catholic church at Langen-Schwalbach, and is also traced on the marble wall of the Capella Emiliana, at San Michele in Isola, near Venice, which was built by Bergamasco in 1530. This latter dial, which has the date 1863, and the reference to Psalm cii. 11, is placed between the statues which occupy the niches of a building that has been stigmatised by Mr. Ruskin as " a beehive set on a low hexagonal tower, with dashes of stonework about its windows, like the flourishes of an idle penman." But apart from the question whether our admiration is due or not to a heavy renaissance chapel, the dial with its motto could not have been more appropriately fixed than beside the shore where the Venetians land their dead for interment in this " quiet sleeping ground in the midst of the sea." In the Church of San Michele, built on the ruins of San Cristoforo della Paçe, Fra Paolo Sarpi is buried.

54.

DIES NOSTRI QUASI UMBRA, ET NULLA EST MORA.

Our days are as it were a shadow, and there is no stay.

Is inscribed on a dial fixed on Tutbury Church in Staffordshire; and is not inappropriate to the place where Queen Mary was lodged, for the last time, on her fatal journey to Fotheringhay.

55.

DISCE BENE VIVERE ET MORI.

Learn to live and die well.

Is on a pillar dial in the churchyard at Conway, which is further inscribed:—"Erected by the Corporation of Conway, Robert Wynne, Jr., Esq., Alderman; Hugh Williams and John Nuttal, Bailiffs, 1761. Meredith Hughes fecit. The difference of 20 places from Conway."

56.

DISCE DIES NUMERARE TUOS.

Learn to number thy days.

Is on an old school-house at Wortley, near Sheffield, and it is also engraved on a large stone dial in the kitchen garden at Barnes Hall, near Sheffield, the property of W. Smith, Esq. The date upon this dial is 1738, and without doubt it was the handiwork of a very remarkable man, Samuel Walker, of Masbro', the father of the Iron Trade of England. He was of humble origin, born in the parish of Ecclesfield, and began life as a parish schoolmaster and dial-maker. When fitting this identical dial at Barnes Hall, then occupied by Sir W. Horton, that gentleman remarked to a friend, "Sam Walker will one day ride in his carriage." The words were prophetic, for in a few years Walker laid the foundation of the largest iron-works in the country at Masbro', near Rotherham, and his descendants have since occupied, and still retain, a good position as county gentlefolk.

57.

DISCITE JUSTITIAM, MONITI.

Learn justice, being warned.

Professor Beckmann, in his "History of Inventions and Discoveries," says: "On the side of New Palace Yard, which is opposite to Westminster Hall, and in the second pediment of the new buildings from the Thames, a dial is inserted with this remarkable motto upon it: *Discite jus-*

titiam, moniti, which seems most clearly to relate to the fine imposed on Radulphus de Hengham being applied to the paying for a clock." The professor proceeds to state that the dial was fixed exactly where Strype describes the clock-house to have stood.

Blackstone tells the well-known story, how Chief Justice Ralph Hengham—" a very learned judge to whom we are obliged for two excellent treatises of practice"—out of mere compassion for a very poor man, altered a fine of 13*s*. 4*d*. to 6*s*. 8*d*., and was consequently fined 800 marks by King Edward I., which were expended in building a clock-house to regulate the sittings of the Courts. This sovereign, who has been styled the Justinian of England, did so much to reform the Courts, that Sir Matthew Hale says, "that more was done in the first thirteen years of his reign to settle and establish the distributive justice of the kingdom, than in all the ages since that time put together." We may consider that the present clock tower at Westminster, from which "Big Ben" gives forth his loud utterances, is a more than sufficient substitute for that with which Judge Hengham's name is associated.

The same motto has been adopted for one of the Temple dials, which was restored in 1861, J. T. A. being the initials upon it.

58.

DOCET UMBRA.

The shadow teaches.

On a large square dial in Austin Friars, London.

59.

DUBIA OMNIBUS, ULTIMA MULTIS. 1835.

Doubtful to all, the last to many.

Is painted on a house at Grasse, in the department of the Alpes Maritimes, France; also on the south wall of the village church at Cambo, Basses

Pyrénées, France, where directly below it is a plain, round-arched window, with two or three cypress trees standing in the churchyard close by. The ground behind the church slopes sharply down to the river Nive, which joins the Adour at Bayonne about fifteen miles off.

Cambo is a purely Basque village at the foot of hills. It has mineral waters, and an *établissement thermale* on the bank of the river, which is much frequented during the summer. Here the peasants hold a fête on the eve of St. John, with dancing and singing of old Basque songs. The history of Cambo may be supposed to commence with Roland the Brave; for not far off is the celebrated *pas de Roland*, or hole, which that hero is said to have kicked through the rock for the purpose of affording a passage for his knights.

Both the dials are modern.

60.

DUM LICET, UTERE.

Whilst it is allowed, use it (time).

Is on a dial in the courtyard of the old Castle of Stazzano, near Serravalle Scrivia, in the province of Alessandria, North Italy. The castle is now a priests' school.

61.

DUM PROFICIT D———T.

This defaced inscription may be seen in the cloisters of Chambery, in Savoy. The reader may amuse himself by supplying the illegible word to his own taste. A friend suggests *deficit*, which seems most probable.

62.

DUM SPECTAS, FUGIO: SIC VITA.

Whilst you are looking, I fly; so does life.

In a three-sided bay-window over a shop in the High Street of Marlborough, is a handsomely illuminated glass dial of oval shape, which nearly occupies four of the twelve panes that compose the projecting centre of the window, and which is inscribed with this motto. A golden scroll on a red ground surrounds the dial face, in the centre of which is a fly so beautifully depicted, that you can hardly believe it is not a real insect incorporated in the glass, as in amber; for it is not perceptible to the touch. There is no gnomon at present; for, singularly enough, it was destroyed by lightning.

At Winchester College there is also the fly in a similar glass dial; and in Leadbetter's book many of the plates of dials have the fly figured. Why is the fly thus introduced? Enquiry has hitherto been made in vain.

The same motto may be read on Ingleton Church, near Settle, in Yorkshire; and in the south wall of the old tower of Willesdon Church, Middlesex, there is a dial, dated 1736, which bears the same inscription, whilst immediately over it has been fixed a valuable chiming clock, the gift of a lady, to which the words equally well apply. This reminds us of the lines in Hood's poem of the "Workhouse Clock," which contain the double allusion:

> "Oh! that the Parish Powers,
> Which regulate Labour's hours,
> The daily amount of human trial,
> Weariness, pain, and self-denial,
> Would turn from the artificial dial
> That striketh ten or eleven,
> And go, for once, by that older one
> That stands in the light of Nature's sun,
> And takes its time from Heaven!"

63.

DUM SPECTAS, FUGIT HORA.

Whilst thou art looking, the hour is flying.

Is on Heighington Church, near Darlington.

64.

DUM SPECTAS, FUGIT HORA: CARPE DIEM.

Whilst you look, the hour flies: seize the opportunity.

These words are on a house-dial at Wolsingham, in the county of Durham. The figures are gilt on a black ground.

65.

DUM TEMPUS HABEMUS, OPEREMUR BONUM.

Whilst we have time, let us do good.

Is one of three dial mottoes that were suggested by the Rev. W. L. Bowles to mark sunrise, noon, and sunset. The author gives the following translation, and applies the couplet to "noon."

> "Life steals away—this hour, O man, is lent thee,
> Patient to work the work of Him who sent thee."

66.

EDI FERRO LO STIL: MA E D'ORO IL TEMPO.

The stile is iron, but time is gold.

May be read at Cambiano in Piedmont. The Dutch have a proverb: "Speaking is silver, silence is gold."

67.

EHEU, FUGACES!

Alas, how fleeting!

A quotation from Horace's Ode, "Eheu fugaces, Postume, Postume, labuntur anni." At Sedbury Hall, near Richmond, Yorkshire, the seat of George Gilpin Brown, Esq., there is a stone pillar-shaped dial attached to the sill of the drawing-room window, with this touching motto upon it.

Also (as Dr. Doran tells us in his "Life of the Rev. Dr. Young"), the author of "Night Thoughts," set up a dial in the rectory garden of Welwyn, Hertfordshire, with the motto, "Eheu, fugaces;" and a few nights afterwards, thieves entered the garden and proved the wisdom of the poet's choice of a motto, by carrying the dial away.

On the walls of the entrance-tower at Farnham Castle, the palace of the Bishop of Winchester, there are two dials which formerly bore the inscription, "Eheu, fugaces, labuntur anni!" Other mottoes, more appropriate to an episcopal residence, have now been substituted, as will be shown hereafter.

68.

ELECTA, UT SOL BEAT ORBEM SPLENDORE.

The chosen one: as the sun blesses the earth with its splendour.

STELLT DEINS LEBENS TAG ZU DIENST MARIA EIN,
SO WIRD DEIN LETZTE STUND IN TOD DIE BESTE SEIN.

Give up the day of thy life to the service of Mary,
So will thy last hour in death be the best.

These mottoes are at Rosenheim, and are evidently laudatory of the Virgin, who is represented as the crowned Queen of Heaven, sitting richly robed on clouds, with the sun's rays behind. A scroll above her has the Latin line, and beneath her feet is a curling double scroll with the figures of the hours above, and the German legend below. The sceptre in the Virgin's hand forms the gnomon. The design is the work of an artist, and makes an elegant picture.

69.

ERRAR PUÒ IL FABBRO,
ERRAR PUÒ IL FERRO,
IO MAI NON ERRO.

The maker may err,
The iron may err,
I never err.

At Graglia in Piedmont. This claim to infallibility on the part of the dial was recently disputed, as shown by a pleasant anecdote which a correspondent has kindly supplied. A clergyman, officiating in the united parishes of Hastingleigh and Elmsted in Kent, as he was entering the church at the latter place, said to the clerk, "What's the time by the dial?" "Well, sir," was the reply, "the dial is half-past ten, but *I think it must be a little fast, as my watch is only ten minutes past ten!*" Both church and churchyard at Elmsted are very interesting, from their antiquity and beauty of situation. The yew trees in the churchyard are especially grand—one of them measures twenty-seven feet in circumference.

70.

ΕΡΧΕΤΑΙ ΓΑΡ ΝΥΞ.

For the night cometh.

A sketch of this dial was made by the collector at Abbotsford in 1839, when the pedestal stood outside a small plantation near the house. But the dial-plate with its gnomon was gone: only two nails which had once served to fasten it remained. So the motto had been a prophecy; for the dial's work was over, since it could henceforth record nothing, except that the night was coming—which indeed had come as if in mockery of itself. One could not help thinking further of the night that came down upon Abbotsford, when its illustrious master was lost to the world.

The motto was also adopted by Dr. Johnson, as we learn from the following passage in Boswell: "At this time I observed upon the dial-plate of his (Dr. Johnson's) watch a short Greek inscription, taken from the New Testament, Νυξ γαρ ερχεται, being the first words of our Saviour's solemn admonition to the improvement of that time which is allowed to us to prepare for eternity—'the night cometh when no man can work.' He some time afterwards laid aside this dial-plate, and when I asked him the reason, he said, 'It might do very well upon a clock which a man keeps in his closet; but to have it upon his watch which he carries about with him, and which is looked at by others, might be censured as ostentatious.'" Croker adds in a note: "the inscription, however, was made unintelligible by the mistake of writing νμξ for νυξ." We would observe that this error is quite sufficient to account for the learned scholar putting aside the watch; and we know that he did not always condescend to fully enlighten his *shadow* "Bozzy," as to his motives. It is also remarkable that in both cases the word γαρ should have been introduced, for it is not in the New Testament. Probably, however, Sir Walter copied the passage from Johnson, without referring to the original.

71.

ESTEEM THE PRECIOUS TIME,
IT MUST PASS SO SWIFT AWAY:
PREPARE THEN FOR ETERNITY,
AND DO NOT MAKE DELAY.

On Wilton Bridge, near Ross, in Herefordshire, there stands a dial, the pedestal of which from its design and enrichment belongs to the Louis Quatorze period, *cir*. 1660—1700. The ball at the top was torn off some years ago, and thrown into the stream; but was afterwards recovered and replaced. It bears the foregoing motto.

72.

ΕΤΙ ΜΙΚΡΟΝ ΧΡΟΝΟΝ ΤΟ ΦΩΣ ΜΕΘ' ΥΜΩΝ ΕΣΤΙ,
ΠΕΡΙΠΑΤΕΙΤΕ ΕΩΣ ΤΟ ΦΩΣ ΕΧΕΤΕ.

YET A LITTLE WHILE IS THE LIGHT WITH YOU,
WALK WHILE YE HAVE THE LIGHT."—*John xii.* 35.

The dial-plate on which these inscriptions appear is fixed on an old school-house at Aynho, near Bicester, now disused. The shape is an oblong square. The sun is represented as a full human face, with jets of light all round. The eyes and mouth are given, and the gnomon forms the nose. " M. C. 1671 " are in the centre, being the initials of the builder and the date of the building. " House built by one Mary Cartwright."

"Should not each dial strike us as we pass,
Portentous, as the written wall which struck,

> O'er midnight bowls, the proud Assyrian pale,
> Erewhile high flush'd with insolence and wine?
> Like that the dial speaks; and points to thee,
> Lorenzo, loth to break thy banquet up!
> 'O man, thy kingdom is departing from thee;
> And while it lasts, is emptier than thy shade!'
> Its silent language such."—YOUNG'S *Night Thoughts*.

73.

EVERY HOUR SHORTENS LIFE.

Was on the church porch at Barnard Castle, until the recent restoration. The dial is now laid by in the church tower. The motto is also on " Turner's Hospital," at Kirkleatham, in Yorkshire, a noble charity founded at his birthplace by Sir William Turner, Lord Mayor of London in 1669. The dial is supposed to have been erected about 100 years afterwards. A second motto is given elsewhere.

74.

EX HIS UNA TIBI.

Of these (hours) one is for thee.

This is on a church in Brittany, placed under the usual radiating sun-face.

75.

EX HOC MOMENTO PENDET ÆTERNITAS.

On this moment hangs eternity.

On an old gable in Lincoln's Inn there is a southern dial thus inscribed, which was restored in 1840, and shows the hours by its gnomon from 6 A.M. to 4 P.M. A newspaper of 1812 informs us that a book was one morning found to have been suspended on the gnomon by the hand of some wag. When taken down, the volume proved to be an old edition of *Practice in Chancery*. The same motto is at Sandhurst, Kent, " W. Hawney, *fecit*, 1720.

Latitude 51.0′ 4″; also on Glasgow Cathedral, together with others elsewhere noted.

> "Time from the church-tower cries to you and me
> Upon this moment hangs eternity:
> The dial's index and the belfry's chime
> To eye and ear confirm this truth of time.
> Prepare to meet it; death will not delay;
> Take then thy Saviour's warning—Watch and pray!"
>
> J. MONTGOMERY.

76.

FELICIBUS BREVIS, MISERIS HORA LONGA.

The hour is short to the happy, long to the miserable.

Is painted on an oval dial on a house wall at Martigny. The inscription is below, and the hour-glass and wings of Time are above. This expresses the same sentiment as No. 10. Shakespeare says, in *As You Like It*, Act iii. Scene 2:—"Time travels in divers paces with divers persons: I'll tell you who time ambles withal, who time trots withal, who time gallops withal, and who he stands still withal."

77.

FERREA VIRGA EST, UMBRATILIS MOTUS.

The rod is of iron, the motion of shadow.

The iron rod is of course the gnomon. The first letter of the last word has been defaced, but "motus" seems to be the most suitable term. It is painted on a large square dial on a wall facing north, in the cloisters of the cathedral at Chambéry.

78.

FESTINAT SUPREMA.

The last hour hastens on.

Was copied in North Italy by Mr. Howard Hopley, and recorded by him in the "Leisure Hour."

79.

FIAT LUX, ET FACTA EST LUX, FACTUMQUE EST VESPERE ET MANE DIES UNUS.

"Let there be light, and there was light: and the evening and the morning were the first day."

At Courmayeur. The motto is the Vulgate version of the passage quoted.

80.

FILI, CONSERVA TEMPUS.

My son, economize your time.

Without date and modern, there is a dial painted yellow and deep red on the wall of a house on the Superga, near Turin, on which this motto appears. It is square in shape, and placed close under the roof. The house stands by the road side, very near the Church, and overlooks a valley of olive trees, with the city of Turin in front, backed by snowy mountains.

The place has an historical interest. The church of the Superga was built by Victor Amadeus II., and covers the spot where he and Prince Eugene stood, on the 6th of September, 1706, surveying the half-ruined city below, which had sustained a four months' siege, and round which the French army still lay encamped. Here the princes concerted their plan of action, and here Victor Amadeus vowed that he would build a church if victory were vouchsafed to his arms. A beacon from the Superga informed the garrison of the junction of the relieving forces; and within twenty-four hours from this time the French army was in full retreat, whilst their conquerors had entered the capital, and were attending a thanksgiving-service in the cathedral amidst the wild rejoicings of the Turinese. "*Or questa volta,*" exclaimed Eugene, as he saw the flying French, "*or questa volta tutta l'Italia è nostra, né altro sangue si verserà per la sua conquista.*" The battle of Turin checked the power of Louis XIV. only less than the victories of Marlborough and Eugene in the north; and so strongly was this felt in England, that several persons made wills in favour of Prince Eugene—notably so, an old maiden lady and a gardener. The Duke of Savoy fulfilled his vow, and built a church on the Superga. It was begun in 1715 and finished in 1731, the same year in which its founder made his vain attempt to regain the throne which he had abdicated, and, failing in his endeavour, was carried prisoner to Moncalieri, where he died the next year, thus sadly ending a brilliant life.

The cost of the building was enormous; even the water had to be brought up on the backs of mules. The vault beneath the church is the burying place of the kings of Sardinia. The first buried there was the founder, Victor Amadeus; and the last was another broken-hearted ex-king, Charles Albert, who died in exile at Oporto in 1849, after a still sadder life. For some years after the building of the church, the royal family used to go there yearly in procession, and return thanks for the deliverance of Turin.

The motto on the Superga is also found at Palermo and Carenna.

81.

FINEM RESPICE.

Consider the end.

The place not identified.

82.

FLOREAT ECCLESIA. 1697. L. 54° 12′.

May the Church flourish.

Is on the church-porch at Kirkby Malzeard, Yorkshire; and the dial is further inscribed, "This dial was given by Mr. W. Buck, minister here in anno 1697." To be read also at Marton cum Grafton, in the same county.

83.

FRONTE CAPILLATA, POST EST OCCASIO CALVA.

Opportunity has hair in front, and is bald behind.

This well-known line is inscribed on an old dial on the school-house at Guilsborough, Northamptonshire. It is quoted from *Distichorum de Moribus*, lib. ii. D. xxv., written by Dionysius Cato, who is supposed to have lived in the time of the Antonines, in the second century. The two lines are :—

> "Rem tibi quam nosces optam dimittere noli;
> Fronte capillatâ, post est occasio calva."

Lord Bacon in his xxi. Essay, *Of Delays*, thus writes :—"For occasion (as it is in the common verse) turneth a bald noddle after she hath presented her locks in front, and no hold taken." "Take Time by the forelock," is a proverb; and the conventional figure of Time represents an old man bald, except a tuft of hair on the crown of his head. Shakespeare recognises the same idea in *All's Well that Ends Well*, Act v. Scene 3 :—

> "Let's take the instant by the forward top;
> For we are old, and on our quick'st decrees
> The inaudible and noiseless foot of time
> Steals ere we can effect them."

84.

FUGIO, FUGE.

I fly—fly.

One of several mottoes on a cross-dial at Elleslie, near Chichester.

85.

FUGIT, DUM ASPICIS.

It flies, whilst you look.

The place unknown.

86.

FUGIT, ET NON RECEDIT TEMPUS.

Time flies, and never comes back.

Appears as a dial and a clock motto at once on the wall of a little court in the Convent della Quiete, near Florence. There is an overhanging roof, and above is suspended a tinkling bell. The convent was originally a royal villa, and received its name,—" La Quiete della Granduchessa Christina,"— from its noble owner. It afterwards became the property of Donna Eleonora Ramirez di Montalvo, the foundress and head of the present existing school.

87.

FUGIT HORA.

The hour flies.

On a square stone dial on an old house, called Moat Hall, near Great Ouseburn, Yorkshire.

88.

FUGIT HORA, ORA.

The hour flies, pray.

This is on a circular dial in a square slab of stone over the porch entrance of Catterick Church, near Richmond, Yorkshire. The face is painted blue, the lettering is gilt, and the gnomon springs from a golden sun which is immediately below the motto.

The Rev. A. J. Scott, D.D., the friend and chaplain of Lord Nelson, who died in his arms at Trafalgar, was vicar of Catterick from 1816 to 1840, and was the father of the compiler of this volume. The same inscription occurs at Gilling Church, a few miles from Catterick.

Most exquisitely does Tennyson touch the three successive chroniclers of time—the hour-glass, dial, and watch, in one of the poems of his "In Memoriam."

> "O days and hours, your work is this,
>   To hold me from my proper place,
>   A little while from his embrace,
> For fuller gain of after bliss:
>
> That out of distance might ensue
>   Desire of nearness doubly sweet;
>   And unto meeting, when we meet,
> Delight a hundredfold accrue,
>
> For every grain of sand that runs,
>   And every span of shade that steals,
>   And every kiss of toothed wheels,
> And all the courses of the suns."

89.

FUGIT HORA, ORA, LABORA.

The hour is flying, pray, work.

May be read in Southgate Street, Gloucester.

90.

FUGIT HORA SINE MORA.

The hour flies without delay.

This may be read at North Wingfield, Derbyshire.

91.

FUGIT IRREPARABILE TEMPUS, 1839.

The time flies which cannot be repaired.

On a round dial, south declining east, which is fixed on the church wall just below the angle of the gable and bell-cot at Vallauris, in France. The name of the village dates from the time of the Romans, *vallis aurea*, "the golden valley," so called from the quantity and quality of its oranges. It is situated in a hollow of the hills, not far from Cannes; but is in no other way remarkable, except for the curiously shaped pottery which is there manufactured.

92.

FUGIT, SI STAS.

It flies, if you stand still.

Not identified.

93.

FUMUS ET UMBRA SUMUS.

We are smoke and shadow.

Mr. Howard Hopley records this as inscribed on a dial affixed to the broad chimney of a farm house in Italy. It could not have been better placed.

94.

GEDENKE DASS DU STERBEN MUSST. 1838.

Remember that thou must die.

There is a dial thus inscribed, painted on the south wall, close under the roof of the church at Ringenberg, near Interlachen. The gnomon is in the centre of an eight-pointed star at the top of the dial, and the motto is on a half-circle below. The church was built on the site and out of the ruins of an old castle, and it stands on a hill which overlooks the little lake of Golzwyl, or Faulensee, between the lakes of Thun and Brientz. The tower of the former castle still remains amongst the trees of the churchyard, and commanded in its day a view over a large portion of the lake of Brientz. The church was transferred to this place from Golzwyl in 1674. It has a venerable appearance, and is covered with ivy.

95.

GIVE GOD THY HEART, THY HOPES, THY GIFTS, THY GOLD: THE DAY WEARS ON; THE TIMES ARE WAXING OLD.

In "Tales of the Village," by the Rev. F. E. Paget, v. i. c. i., "The Miser's Heir," contains the following passage: "As I proceeded leisurely round Baggesden Hall, I observed an ancient sun-dial, adorned with heraldic devices, and grotesque emblems of mortality, carved in stone, according to the style which prevailed at the close of the 16th century. On a scroll above it was inscribed 'Homfrie and Elianor Bagges. A.D. 1598;' and beneath it, in smaller but still very legible characters, the following rhyme." We have authority for stating that this is altogether imaginary; but the couplet is too pretty not to be retained.

96.

GO ABOUT YOUR BUSINESS.

Is on the church at Kilnwick-on-the-wolds, Yorkshire. The same sentiment is expressed by Nos. 11 and 22.

97.

HANC QUAM TU GAUDENS IN GNOMONE CONSULIS HORAM, FORSITAN INTERITUS CRAS ERIT HORA TUI.

This hour which thou in thy joy consultest on the gnomon.
Perhaps will be to-morrow the hour of thy death.

Is on a house dial at Voltri, near Genoa.

98

HASTE, TRAVELLER, THE SUN IS SINKING LOW;
HE SHALL RETURN AGAIN, BUT NEVER THOU.

Recorded by Mr. Howard Hopley.

99.

῾Η ΣΚΙΑ ΚΟΥΦΗ ΣΟΦΙΑΝ ΣΕ ΔΙΔΑΣΚΕΤΩ.

Let the slight shadow teach thee wisdom.

At Torrington, Devon.

100.

HERE MI, NESCIS HORA
MORIERIS, SI QUÆRIS, QUA. John Owen, 1683.

My master, you know not the hour in which you will die, even if you ask.

A bad Latin motto on a spirally-carved stone dial in the garden of the Hon. W. O. Stanley, at Penrhos, Holyhead. Let no one imagine that this motto is either misspelt or mistranscribed. Mrs. Vaughan, from whom the collector received it, with a drawing of the dial, vouches for its accuracy on the authority of the Master of the Temple, who compared it with the original inscription, and found it correctly copied, however horribly incorrect in itself. It afterwards brought a smile to the lips of the Laureate, who, when translating it broke out into exclamations, "But you've no notion what bad Latin it is! But you can't imagine how vile the Latin is!" *Oh! my master, if thou shouldest seek to know the hour of thy death, thou shalt be ignorant of it.* We have added the word "even" in our translation, in order to make the idea a little more intelligible.

101.

HEU, PATIMUR UMBRAM.

Alas! we endure the shade.

On a dial at Sleningford Hall, near Ripon, the seat of the Dalton family. It may be feared it is no longer there. A friend of the collector had a search made for it a few years ago, in order to let her know whether it was on the house, or in what part of the grounds. The result was unfortunate: it could not be found at all.

.102.

HEU, QUÆRIMUS UMBRAM.

Alas! we pursue a shadow.

Recorded by Mr. Howard Hopley.

103.

HIC LICET INDULGERE GENIO.

Here you may indulge your taste—do what you like.

"Indulgere genio" is a phrase from Persius. Dean Alford says, "I observed between Mentone and Bordighera, a brand new villa conspicuously inscribed," as above. "On inquiry I found that it belonged to an eccentric lady."

104.

HINC —— DISCE ——

In Malvern churchyard stands a graceful shaft, nineteen feet high, of a mortuary cross of the fifteenth century, crowned with a cube and ball, the latter being probably an addition of the last century. The dial on the sides of the cube presents four faces to the points of the compass. On the north face there is an illegible inscription. Above are the initials W. K., most likely those of the maker, with two or three defaced words, one apparently being *hinc*, and below are three words, that in the centre being *disce*. Half way up the shaft is a pretty niche, in which no doubt was a figure of the Virgin.

105.

'Ο ΚΑΙΡΟΣ ΟΞΥΣ.

Time is swift.

'ΑΜΕΡΑΙ ΕΠΙΛΟΙΠΟΙ ΜΑΡΤΥΡΕΣ ΣΟΦΩΤΑΤΟΙ.

The soundest testimony is borne by the days that remain to us.

Is on a dial on Alleyne's Grammar School, at Uttoxeter. It was originally on the old school-house that was built in 1568, and was removed to the new one erected in 1859.

106.

HORA BIBENDI.

The hour of drinking.

Is a dial motto which is aptly used as the sign of a public-house, near Grenoble.

.107.

HORA FLUIT, CULPÆ CRESCUNT, MORS IMMINET:
HEU, VITÆ CORRIGE FACTA TUÆ.

The hour flows on, our faults increase, death impends : alas, correct the deeds of thy life.

This dial motto is on the church of St. Pierre, in Switzerland, and the words are illegibly traced. There is no date; but age or the tempests of those high regions have defaced the inscription. St. Pierre is on the road to the pass of the great St. Bernard, 5,302 (Swiss) feet above the level of the sea. It is a wild and dreary looking village. Dark rough chalets, raised a foot or two from the ground, are scattered about the neighbouring fields. The river roars and foams in the deep defile below; and pine trees, the remains of the forest through which Napoleon's army struggled with so much difficulty in 1800, cover in patches the mountain sides.

The dial is painted on the church wall protected by a buttress, and facing the snowy Mont Velan. The church dates from 1018, and a Latin inscription, placed by its founder, Bishop Hugo of Geneva, commemorates a Saracen invasion. A Roman military column, dedicated to the younger Constantine, and said to have replaced a statue of Jupiter on the summit of the Pass, but destroyed by Constantine about 339 A.D., attests the great antiquity of the place.

108.

HORA, DIES, ET VITA FUGIT, MANET UNICA VIRTUS,

The hour, day and life, all fly away: virtue alone remains.

Is noted in Cyrus Redding's "Fifty Years' Recollections, Literary and Personal," vol. iii. p. 86.

109.

hora horIs CEdIt, pereVnt sIC teMpora nobIs :
Vt tIbI fInalIs sIt bona, VIVe bene.

An hour yields to hours, so our time perishes:
That thy last hour may be good, live well.

Many years ago the collector's old and kind friend, the late Lord Chief Justice Tindal, brought over for her from Karlsbad a mysterious inscription, which he had carefully copied in scholarly handwriting. The dial was formed on two sides of the angle of the upper story of a substantial house in the market-place. The Chief Justice wrote, " The letters which are written in capitals were so in the original inscription, and were coloured red: probably the anagram of some one's name is concealed under them." By consulting that useful oracle, " Notes and Queries," we had the difficulty solved. We suggested that it might be a chronogram, but for the introduction of the letter E. A correspondent replied that probably CEdIt ought to be written CeDIt, when the following numerals could be extracted: MDCCVVVVIIIIIIIIII : MDCCXXX : 1730, which we may suppose to be the date of the building.

It is amusing to record further, that some friends who were staying more recently at Karlsbad, kindly looked after this dial, which they found, but in a dilapidated state. They made out the motto, however, with the help of the Burgomaster of the place, who owned that he had lived opposite to it all his life, but had never noticed it. Nevertheless, he became much interested, and said he would give orders that it should be cleaned and repainted. The Doctor, too, confessed that he had never seen it before, but should henceforth point it out to his patients for their contemplation and improvement.

110.

HORA PARS VITÆ.

An hour is a portion of life.

Fixed to what appears to have originally been a mortuary cross in Brading Churchyard, Isle of Wight, is a metal dial-plate dated 1815, with the foregoing motto. "J. James, G. Hearn, Churchwardens. J. Wood, *fecit.*" The stone column is about four feet high, rising from three circular steps, which are much worn, and, like the shaft, appear to be of considerable age. As the date 1715 also appears, this will probably refer to a previous dial, from which the motto may have been borrowed.

111.

HORA PARS VITÆ, HORA PARS UMBRÆ,

The hour is a portion of life, the hour is a portion of shade,

Is on a pillar-dial in the old churchyard of Castleton, in Derbyshire.

112.

HORA NOBIS LUCRUM.

An hour is a gain to us.

An iron-plate, probably belonging to a former sun-dial, and bearing this motto, is on the porch of Huish Episcopi Church, near Langport, Somersetshire.

113.

HORÆ PEREUNT ET IMPUTANTUR.

The hours perish and are placed to (our) account.

Occurs on the Riviera.

114.

HORAM SOLE NOLENTE NEGO.

I refuse to tell the hour, when the sun is unwilling to shine.

May be read at Poirino, in Piedmont.

.115.

HORAS NON NUMERO NISI SERENAS.
I only reckon the bright hours.

This elegant motto is on dials at Sackville College, East Grinstead; on the Town Hall at Aldeburgh in Suffolk, which was built *circa* 1500, and has been recently restored; at Leam, near Leamington; in the garden of Beard Sheppard, Esq., Frome, having been removed from the rectory garden of Compton Basset, Wilts; in front of a farmhouse near Farnworth, Lancashire; at Campo Dolcino; at Cawder, near Glasgow; at Arley Hall, Cheshire; and, according to Hazlitt, near Venice. The motto is too good to be uncommon, and is thus alluded to by Mr. Helps in his "Friends in Council," First Series, vol. I. book II.: "Milverton had put up a sun-dial in the centre of the lawn, with the motto, 'Horas non numero nisi serenas;' which gave occasion to Ellesmere to say, that for men the dial was either totally useless or utterly false."

We cannot resist giving the well-known testimony of Shakespeare in favour of the shepherd's life for serenity and peace. Henry VI. thus soliloquizes on the battle-field at Towton:—

> "O God! methinks, it were a happy life,
> To be no better than a homely swain;
> To sit upon a hill, as I do now,
> To carve out dials quaintly, point by point,
> Thereby to see the minutes how they run:
> How many make the hour full complete,
> How many hours bring about the day,
> How many days will finish up the year,
> How many years a mortal man may live.
> When this is known, then to divide the times:
> So many hours must I tend my flock;
> So many hours must I take my rest;
> So many hours must I contemplate;
> So many hours must I sport myself;
> So many days my ewes have been with young;
> So many weeks ere the poor fools will yean;
> So many years ere I shall shear the fleece:
> So minutes, hours, days, weeks, months, and years,
> Pass'd over to the end they were created,
> Would bring white hairs unto a quiet grave."

116.

HORTUS UTRAMQUE TULIT, NOS ET MEDITEMUR IN HORTO.

The garden bore both, let us also meditate in the garden.

This motto is on a dial in the Nuns' Garden at Polesworth, near Tamworth. It must be imperfect; and it has been suggested that a previous line may have referred to the two trees of Life and Knowledge in the Garden of Eden. If so, the meaning is clear.

Peculiar interest attaches to the foundation of the Benedictine nunnery at Polesworth. Dugdale gives the following account: "Egbert, king of the West Saxons, built this monastery of nuns, and made his daughter Edith the first abbess, having caused her to be instructed in the Rule of St. Benedict by Modwen, an Irish lady, whom he had sent for out of that country, because she had there cured his son, Arnulf, by her prayers, of a leprosy. King William the Conqueror gave to Sir Robert Marmyon the castle of Tamworth, with all the lands about it, in which was the nunnery of Polesworth. This knight turned out the nuns; but a year after, being terrified by a vision, he restored them, they having retired during that time to a cell they had at Oldbury or Aldbury, given to their monastery by Walter de Hastings. However, the aforesaid Marmyon was afterwards reckoned the founder of Polesworth."

This spot appears to have been the site of the first religious house that was planted in the centre of England, and one of the first that found a local habitation in the kingdom. The name of the foundress is still preserved in the neighbourhood. The parish church of Burton-upon-Trent is dedicated to the joint names of St. Mary and St. Modwenna. The site of her chapel is still called "St. Modwen's Orchard," and "St. Modwen's Well" was celebrated, two hundred years ago, for the sanatory properties of its water. The nunnery became the place of education to which the young ladies of the highest families were sent before they entered the society of the world.

The nunnery was dissolved in 1539, when Sir Francis Nethersole became possessed of the conventual lands, and built the hall out of the ruins of the nunnery. It is supposed that the dial was then created in the centre of a square garden on the site of the cloisters. It is now placed on the corner of an old wall, as if to get it out of the way. The garden has disappeared, but the spot is still an orchard with a pretty green sloping to the river side. As to the construction of the dial: there is a projecting base surmounted by several courses of wall stone on which is the principal object. This consists of a curved pediment of stone, supporting a square block on the east side of which is represented a tomb: below is the motto, and on a scroll above are the words, "Non est hic: resurrexit"—*He is not here: he has risen.* The top is finished off so as to correspond with the pediment, and contains the Nethersole coat of arms. Among the devices are the Death's head and cross bones: also an apple, which seems to identify the reference in the motto with—

> "The fruit
> Of that forbidden tree, whose mortal taste,
> Brought death into the world, and all our woe."

117.

HORULA DUM QUOTA SIT QUÆRITUR, HORA FUGIT. 1678.

Whilst you ask what the little hour is, the hour flies by.

In the churchyard of Kirk-Arbory, Isle of Man, is a dial-plate thus inscribed. Above is "Thomas Kirkale de Bolton fecit," and below is a coat of arms.

118.

I ALSO AM UNDER AUTHORITY.

As this motto has been published in the pages of "Aunt Judy's Magazine," it may claim to take its place with those culled from the writings of Wordsworth and Mr. Paget; and all of them may be regarded as suggestions for new dial inscriptions.

119.

I AM A SHADOW, SO ART THOU:
I MARK TIME, DOST THOU?

Is inscribed on a sun-dial in Stirling Castle Cemetery.

120.

I ONLY MARK BRIGHT HOURS.

There is a square stone pedestal, panelled on the four sides, and surmounted by a dial which bears this motto; and it was so inscribed by the late Countess of Tyrconnel, in whose garden at Kiplin Hall, near Catterick, Yorkshire, it was erected.

> "For as in sunshine only we can read
> The march of minutes on the dial's face,
> So in the shadows of this lonely place
> There is no love, and Time is dead indeed."
>
> *Sonnet by Hood.*

121.

I STAND AMID Y^E SUMMERE FLOWERS
TO TELL Y^E PASSAGE OF Y^E HOURES.
WHEN WINTER STEALS Y^E FLOWERS AWAYE
I TELL Y^E PASSING OF THEIR DAYE.
O MAN WHOSE FLESH IS BUT AS GRASSE
LIKE SUMMERE FLOWERS THY LIFE SHALL PASSE.
WHILES TYME IS THINE LAYE UP IN STORE
AND THOU SHALT LIVE FOR EVER MORE.

Sent to the collector by her friend, the Rev. Greville J. Chester, as being inscribed on the four sides of a dial in a flower-garden at S. Windleham. It was an ingenious practical joke, as the lines were invented for the occasion; but they are so pretty and quaint, that she is loath to let them pass away unrecorded.

122.

IF O'ER THE DIAL GLIDES A SHADE, REDEEM
THE TIME; FOR, LO, IT PASSES LIKE A DREAM.
BUT IF 'TIS ALL A BLANK, THEN MARK THE LOSS
OF HOURS UNBLEST BY SHADOWS FROM THE CROSS.

This wise monition is on a dial which bears the form of a cross in a leaning position, at the top of a pillar in Shenstone Churchyard, near Lichfield. The hours are marked on the sides of the shaft, and the shadows are cast on the figures by the arms of the cross. It was erected a few years ago by a lady. The same motto is on a cross dial at the Church of S. Mary the Virgin, Collaton, Devonshire. A fine cross-dial was erected more than thirty years ago in the Rectory garden at East Lavant, Chichester, by the Rev. H. Legge, but it is not inscribed. A similar dial, without a motto, is in the gardens of J. Guest, Esq., Rotherham, which tells the time of day at Paris, Geneva, New York, Calcutta and Pekin. It seems to be cut in Roche Abbey stone, and the weather is affecting it. "J. Booth fecit. Latitude 53·24. Long. 1 27."

123.

IMPROVE THE TIME. 1765.

Appears on an oval dial on the Unicorn Inn at Uppingham.

124.

IMPROVE THE PRESENT HOUR, FOR ALL BESIDE
IS A MERE FEATHER ON A TORRENT'S TIDE.

May be read at Fredericton, in New Brunswick, Canada, as the motto of a dial, which is placed on a wooden shaft that is shaped like a ninepin, and stands in the garden of the late Mrs. Shore.

125.

IN SUCH AN HOUR AS YE LOOK NOT FOR, THE SON OF MAN COMETH. 1796.

Is on a plain oval dial erected against the wall of Bakewell Church, Derbyshire, nearly under the roof.

126.

INCESSANT DOWN THE STREAM OF TIME, AND HOURS, AND YEARS, AND AGES ROLL.

On the plate of a pillar-dial in the kitchen-garden at Lansdowne Lodge, Kenmare, County Kerry.

127.

INDUCE ANIMUM SAPIENTEM. 1775.

Put on a wise mind.

May be seen on the south porch of the Church in the pretty retired village of Eyam, in Derbyshire; noted for a fearful visitation of the plague, which nearly depopulated the place, in 1666. The dial is very elaborate, and has the tropics of Cancer and Capricorn, &c. marked upon it; also the names of places, showing their difference from English time. The names of "Wm. Lee, Tho. Froggatt, Churchwardens," are recorded.

128.

IN LUCEM OMNIA VANA.

All things are vanity (when brought) to the light.

A plain south dial on the new hydropathic bath-house at Biella-alta, in Piedmont, bears this motto.

129.

THE HOURE OF DEATHE D BE MERCIFUL UNTO ME.

HE THAT WILL THRIVE
MUSTE RISE AT FIVE;
HE THAT HATH THRIVEN
MAY LIE TILL SEVEN;
HE THAT WILL NEVER THRIVE
MAY LIE TILL ELEVEN.

FOR AS TYME DOTH HASTE SO LIFE DOTH WASTE.

In front of Stanwardine Hall, near Baschurch, Shropshire, a fine old Elizabethan mansion, now converted into a farmhouse, stands a pillar-dial, having a silver plate, on which these homely maxims are engraved. The face of the dial is circular, but drawn in a square; and the four vacant corners of the square are occupied above by the two couplets we have given; and below, on one side, by an elephant with a castle on his back, and in the opposite corner is a squirrel sitting up and eating. The advice for early rising is in the centre of the face; beneath is a bird on a shield, and lower down "anno 1560." Stanwardine Hall belonged to the Corbet family, from which it passed to the Wynnes; and it is now in other hands. The elephant and squirrel are the Corbet crests.

130.

IO VADO E VENGO OGNI GIORNO
MA TU ANDRAI SENZA RITORNO.

I go and come every day,
But thou shalt go without returning.

This fine sentiment is on the wall of a modern house in the Rue de France, Nice. At the side of the inscription is the dial, a large semi-oval in shape. It is sketched in brown on the white-washed wall. The street is long and poor-looking, and before the days of railway travelling, was the tourist's entrance into the town from France. The late Dean Alford thus translates the motto:—

"I come and go, and go and come, each day;
But thou without return shalt pass away."

Dr. Henry, in his published Poems, gives the following renderings:—

TRANSLATION.
"I daily come and daily go,
But thou, once gone, com'st never mo'."

PARAPHRASE.
"I'm daily born and daily die,
Thou'rt born but once, but once to die,
And there's an end. Be off! good bye,
Poor silly fool!—great Time am I."

At Arma on the Riviera, and also at Pisa the same motto may be seen; and in the pleasant garden of Monkton Farleigh, near Bath, it is engraved on a pillar dial, but disfigured by mis-spelling.

131.

IRREVOCABILIS HORA. 1842.

The hour which cannot be recalled.

The Bridge over the Siagne, not very far from Cannes, is crossed by the high road from Toulon to Nice; and on the eastern side of the river there is a small house for the toll-keeper, facing south, with an almost round dial painted white on its yellow wall, and super-inscribed as above. Below is the garden window of the *garde du pont*, with its green shutters to keep out the strong sunshine. Beyond the corner of the house are olive-covered plains, from which the hills rise suddenly, and the glowing range of the Alpes Provençeaux forms the background. The same motto was found in the neighbourhood, badly written and mis-spelt, dated 1850, over a south dial on the door of a shed in the valley of Gourdalou, two miles from Cannes. The shed was a roughly plastered building, standing in the midst of heath, which in some places rose to six feet in height, and sheltered by pine trees that had happily escaped the axe which laid bare the surrounding hills.

132.

ISTA VELUT TACITO CURSU DILABITUR UMBRA,
TRANSIT IN ÆTERNOS SIC TUA VITA DIES.

As that shadow glides away with silent step,
So thy life passes into the days of eternity.

This was read somewhere in Tuscany.

133.

ITALICUM SIGNAT TEMPORA SACRA DEO.

The italic (numeral) marks the times that are sacred to God.

On St. Peter's at Rome. Probably the canonical hours of prayer are marked on the dial face in *italic* numerals.

134.

J'AVANCE.

I go forward.

This dial motto is in the garden at Hall Place, near Maidenhead, Berkshire. It is the family motto of Sir Gilbert East, Bart., whose crest is a horse, passant sab.

135.

JAM PROPERA, NEC TE VENTURAS DIFFER IN HORAS,
QUI NON EST HODIE, CRAS MINUS APTUS ERIT.

Hasten on at once, nor delay to future hours:
He who is not ready to-day will be less so to-morrow.

On a house near Newton House Woods, near Whitby. The first line is above, the second below the dial face.

136.

JE LUIS POUR TOUT LE MONDE,
MON OMBRE PASSE AVEC VITESSE,
ET TA FIN APPROCHE AVEC RAPIDITÉ, O MORTEL.

I shine for the whole world,
My Shadow passes on swiftly,
And thy end rapidly approaches, O Mortal.

The sun's full face, surrounded by rays, forms the dial, the gnomon projecting from the centre. Below the solar portrait is the first line: the other two being written under the dial plane. The initials of the maker, and date, F. M. 1833, are on the face. It is placed over the door of a village inn at Rougemont, in the Canton de Vaud, Switzerland, one of the picturesque wooded châlets of the country, and there is an elaborate cornice supporting the deep over-hanging roof. The windows, fitted with little square panes of glass, are small, and flanked by solid shutters; and probably only one pane in the whole house can be opened to admit the outer air, which is so unwelcome to the Swiss domestic hearth. Below is a second cornice over a round arched porch, with thick heavy doors; where may be seen the hostess on a seat knitting, in her black silk cap and lace lappets, with her beehive hat hanging on the wall beside. Such are the last remains of the old costume of the Canton, which, though dying out on the shores of the Lake of Geneva, still linger in the high valleys, where fields creep up to the foot of the precipices, and rocks uprear their sharp peaks against the sky—where too, as Mr. Ruskin boldly says, "the pine forests cover the mountains like the shadow of God, and the great rivers move like His eternity."

Rougemont stands on the high road leading from the Lake of Geneva to Thun, through the Simmenthal. It is the last village in the Canton de Vaud, and the Canton of Berne is hardly a mile off. The language of the people changes from French to German with the boundary. Rougemont is in the "Pays d'en haut Romand," and belonged in old times to the Counts de Gruyère, and was sold by them, together with Gessenây, to the Bernese for 22,000 crowns in the sixteenth century; but unlike its neighbour Gessenay, (now Saanen) it has kept its language and nationality. The first book printed in the Pays de Vaud by the Monks of Rougemont was a Latin Bible, before 1534. The convent is now a chateau.

137.

JE VIS DE TA PRESENCE, ET MON UTILITÉ FINIT EN TON ABSENCE.

I live by thy presence, and my usefulness ends in thy absence.

This is the last half of a motto, near Courmayeur. As some words at the beginning have been defaced, it is impossible to give the meaning of the whole inscription; but it was probably an address to the sun.

138.

ΚΑΙΡΟΝ ΓΝΩΘΙ.

Know the proper season.

Is over a dial placed on a buttress on the north side of Ely Cathedral. The face of the sun seems to emit the lines to the surrounding figures, as well as the gnomon; and between the lines are the signs of the zodiac.

139.

LA VIE DE L'HOMME PASSE COMME L'OMBRE.

The life of man passes as a shadow.

At Courmayeur.

140.

LABUNTUR ANNI.

The years glide away.

Formerly on one of the two dials which were on the entrance tower at Farnham Castle.

141.

LE TEMPS PASSE, L'ETERNITÉ S'AVANCE.

Time passes, eternity approaches.

At Entrèves, near Courmayeur; where there is a second line, so much defaced that no sense can be made of it.

142.

LET NOT THE SUN GO DOWN UPON YOUR WRATH.

Is St. Paul's monition (Ep. iv. 26), and was admirably used by Bishop Coplestone for a dial motto in a village near which he resided.

143.

LEX DEI, LUX DIEI:
LUX UMBRA DEI. 1809.

The Law of God is the light of day;
Light is the shadow of God.

This inscription, together with "Qualis vita," &c. see No. 231, is on the church at Great Smeaton, near Northallerton. It is also to be seen at Rugby.

144.

LIFE'S BUT A WALKING SHADOW. 1769.

From *Macbeth*, Act v. Scene 5, is on a dial facing to the south of an old house in Salisbury Close, formerly inhabited by James Harris, the author of "Hermes," a Salisbury man; who, as he died in 1780, may have erected and inscribed the dial.

145.

LIFE IS SHORT, TIME IS SWIFT,
MUCH IS TO BE DONE. J. S. 1833. Lat. 54, 30.

The dial-plate is circular and made of slate, and is erected on a barn near Bassenthwaite, Cumberland.

146.

LOQUOR, SED NON CÆCIS.

I speak, but not to the blind.

In Mrs. Schimmelpenninck's account of her visit to the ruins of Port Royal, she states that in the burying-ground attached to the chapel, dedicated to the Blessed Virgin, there was a sun-dial which bore this motto: and she adds, " above the portal entrance to the burying-ground were the following inscriptions:—without,

Time is yet before thee;

within,

Time is for ever behind thee.

A quaint verse in old French was also often repeated:—

Tous ces morts ont vécu, toi qui vis, tu mourras:
Ce jour terrible approche, et tu n'y penses pas.

which might be thus rendered:

These dead once lived, and thou who liv'st shalt die:
Thou heed'st it not, yet that dread day draws nigh."

147.

L'ORA CHE L'OMBRA MIA FEDEL TI ADDITA,
PENSA CHE FU SOTTRATTA ALLA TUA VITA. 1862.

The hour which my faithful shadow points out to thee,
Think that it has been withdrawn from thy life.

This together, with "Amicis quælibet hora," is on a house at Murano, near Venice.

148.

LORSQUE TU SONNERAS, JE CHANTE.

When thou shalt strike, I crow.

On a house in the Rue d'Antibes, at Cannes, is a circular dial, surmounted by a gaily-feathered cock. Right and left of the bird spreads a scroll, which is thus inscribed. It is of course the cock's challenge to the dial. The dial faces south, and is green in colour. No date appears, but the once brilliant plumage of the cock, and the condition of the dial face, testify that they have long passed their prime.

149.

L'OROLOGIO PUÓ ERRAR SEGNANDO LE ORE,
MA LA SFERA DEL SOLE GIAMMAI TRASCORRE.

The clock may err in telling the hour,
But the orb of the sun never diverges.

Is on a mill near Riva, Lago di Garda.

150.

LUX POST UMBRAM.

Light after shadow.

May be read as a dial motto in the north of Italy.

151.

LUX UMBRA DEI.

Light is the shadow of God.

From the north of Italy.

152.

1740 Years of

ממשיר

A Stone of Stumbling.
See Isaiah viii. 14, 15.
Ps. cxix. 165. Ezek. iii. 20.
A Stumbling Block.
Beware of Him.
Mal. i. 11.
Bezaleel Benevent
Sculptor Israelite. Isaiah, xliv. 5.
Maker. I am 58 years old.

This extraordinary inscription is carved in stone on the two sides of a dial plate which is inserted in the slab and fixed against a house in the village of Wentworth, on Earl Fitzwilliam's Yorkshire estate. It has puzzled many passers by; but the Rev. Dr. Moses Margoliouth has offered a solution of the mysterious motto in "Notes and Queries," 1st Ser. vol. iv. p. 378. He assumes it to have been the work of a Jewish mason, probably employed in the erection of Wentworth Woodhouse, who had become a convert to Christianity, and who sought to allure his Hebrew brethren to a like change of faith. The Hebrew characters form no word that can be found in the language, but they are the initial letters of the following words:—

מלך משיח שילה יהוה רעי

which express, "The King Messiah, the Shiloh, the Lord my Shepherd." Dr. Margoliouth regards the motto as a veiled admission on the part of the Israelite of his conversion to Christianity, given after a national mode of Eastern communication. It will be observed that the Scriptural references are confined to the books of the Old Testament, so as not to alarm the inquiring reader. Dr. Margoliouth concludes his criticism thus:—"One may well imagine an Israelite or two observing from the road the Hebrew characters ממשיר for they are very large, and are seen afar off—and after puzzling over their intent and purport for some time, proceed to ask for an explanation from the major-domo. The master, delighted that the bait caught, vouchsafes, in his peculiarly eccentric style, to lecture on his own device, and thus reads to his brethren a sermon in stone." By referring to the passages cited in the inscription, the reader will better understand the learned Hebraist's interpretation.

153.

MACHINA, QUÆ BIS SEXTAS JUSTE DIVIDIT HORAS, JUSTITIAM SERVARE MONET, LEGESQUE TUERI.

The instrument, which exactly divides the twelve hours,
Warns you to maintain justice, and guard the laws.

This appears near Notre Dame, Paris.

154.

MÆSTIS LENTÆ, CELERES GAUDENTIBUS HORÆ.

Slow move the hours to the sad, swift to the joyful.

At Stra, near Padua.

155.

MAN FLEETH AS A SHADOW.

A square dial, painted red with a green border, is on a gable over the porch of the picturesque old church at Wycliffe on the Tees. Wycliffe is the reputed birthplace of the Great Reformer, and is very beautifully situated. The same motto was on the church porch at Staindrop, county Durham.

156.

MANE PIGER STERTIS, FUGIT HORA.

In the morning you sluggish snore—the hour flies.

Adapted from Persius, and recorded as a dial motto, but no locality assigned.

157.

MANEO NEMINI.

I wait for no one.

There is a dial which bears this inscription, surrounded by creeper foliage, on Middleton Tyas Hall, near Richmond, Yorkshire.

158.

MAY THE DREAD BOOK AT OUR LAST TRIAL,
WHEN OPEN SPREAD, BE LIKE THIS DIAL;
MAY HEAVEN FORBEAR TO MARK THEREIN
THE HOURS MADE DARK BY DEEDS OF SIN;
THOSE ONLY IN THAT RECORD WRITE
WHICH VIRTUE, LIKE THE SUN, MAKES BRIGHT.

On a dial which projects from the sill of the library window at Arley Hall, Cheshire, the seat of R. Egerton-Warburton, Esq. It has also "Horas non numero nisi serenas."

159.

MEAM NON TUAM NOSCIS.

You know my time, not your own.

At Poirino, in Piedmont.

160.

ME LUMEN VOS UMBRA REGIT.

The light directs me, the shadow you.

At Lesneven, in Brittany.

161.

MEMENTO HORÆ NOVISSIMÆ. 1798.

Remember thy last hour.

This is inscribed on a semi-circular dial on a low cottage or shed beside the Cornice road, on the eastern side of Bordighera. It is placed almost immediately under the roof, the motto and date being below. On the right-hand side a lamp projects from the wall, and hangs in front of a niche, where there is an image of the Madonna. This figure in dress and appearance resembles those which are commonly found in way-side shrines.

162.

MEMENTO MORI.

Remember, you have to die.

Is on a circular dial fixed in a sort of hatchment framing on the wall of Croft Church, in Yorkshire, and half concealing a window. The motto is above, the date 1816 on the sides, and the numerals and lettering are gilt. The same motto, dated 1804, is at Monthey, Canton Vallais, in Switzerland.

"Forget not death, O man, for thou may'st be
Of one thing certain—he forgets not thee!"

A correspondent reminds us that a certain well-known Fellow of Worcester College, Oxford, suggested as the motto for a snuff-box made out of an old mulberry tree, "memento mori"—*remember the mulberry tree.*

163.

MEMOR ESTO BREVIS ÆVI. 1764.

Be mindful of the shortness of life.

There is a circular dial placed over the porch entrance at Bittadon Church, North Devon, which contains this inscription. The whole porch is enveloped in ivy, from which the dial face peers out.

164.

A. D. S.
MIA VITA É IL SOL: DELL' UOM LA VITA E DIO,
SENZA ESSO É L' UOM, QUAL SENZA SOL SON' IO.

My life is the Sun: God is the life of man;
Man without Him, is as I am without the sun.

On the wall of a monastery, now suppressed, in the neighbourhood of Florence.

165.

MI FECE D' ARCHIMEDE L' ALTA SCUOLA,
IL SOL MI DA LA VITA E LA PAROLA.

The lofty teachings of Archimedes made me,
The sun gives me life and speech.

The dial plate is oval in shape, but wider than it is long. The motto is on the top, the points of the compass are marked in the centre, and the date, 1859, is below, with the name of the maker: *Carolus Sachi, Trigon, Desine, Pinxit*. It is erected on the wall of one of the courts of the immense chateau of the Counts Arconati, at Rho, near Milan. The last descendant of this ancient Milanese family, which dated from the fourteenth century, was buried at Milan in 1870.

166.

MONEO, DUM MOVEO.

I warn whilst I move.

On a summer-house at Danby Hall, near Leyburn, Yorkshire.

167.

MONEO, DUM MOVEO:
DUM SPECTAS FUGIO:
SAPIENTIS EST NUMERARE:
SIC PRÆTERIT ÆTAS.

I admonish whilst I move: Whilst you look I fly: It is the part of a wise man to number (his days): So time passes away.

In the private diary of a gentleman, dated 1790, has been found an entry of an interesting group of four mottoes, no longer existing at King's Lynn, Norfolk. "The market-place at Lynn very fine and spacious; a very fine Market Cross, as it is called—a very elegant building, standing on pillars, adorned with statues and four dials, on which are the four mottoes"—as written above. It appears that this interesting structure was built in 1710, and having become dilapidated was replaced by a market house.

168.

MORS DE DIE ACCELERAT. 1796.

Death hastens on day by day.

This inscription was on a dial over an archway in the stable-yard at Kiplin Hall, near Catterick.

When the collector last saw it, in 1864, the motto had been painted over. The dial was made by a villager named Bonner, who died about 1818; and in 1838 the collector sketched his widow at her cottage in Kiplin, and received the information from her.

The same motto is in the churchyard at Derwent, in Derbyshire. This dial is made of a soft grey stone or slate, in shape like an heraldic shield, and is mounted on an oak beam, which was probably taken out of the old chapel of the fourteenth century.

169.

MOX NOX.

Soon (comes) night.

Is a dial motto at Elsworth, near Cambridge. It also occurs on a flint-built church, near Dennington, Suffolk, where the dial is fixed on the battlement, and beneath, on a scroll is:—

> Mox Nox.
> THE MOMENT PAST,
> LAID MANY FAST.

Many of these flint-built churches are very handsome. Round the base of this one the flints are arranged in patterns, to represent the emblems of the Passion, &c., &c.

170.

MULIER, AMICTA SOLE,
ORA PRO NOBIS, SANCTA DEI GENITOR.

Lady, clothed by the sun, holy mother of God, pray for us.

These lines are to be read on the house of the Roman Catholic Priest at Hallstadt, near Salzburg. Evidently new and roughly painted on the wall is the figure of the Virgin holding the gnomon of a dial, which casts its shadow on a scroll beneath, on which the hours are figured.

171.

MY DAYS ARE LIKE A SHADOW THAT DECLINETH.

From Psalm cii. 11, is inscribed on the plate of a pillar dial, formed by a cluster of three slight columns at Haley Hill Cemetery, near Halifax. There is no date, but the dial was probably erected in 1856, when the cemetery was opened. The spire of the fine church, built by Mr. Akroyd, and the chimney of his mills, rise in the background, above the trees which surround the enclosure.

172.

NASCE, MUORE.

It is born, it dies.

At Dolce Acqua on the Riviera.

173.

NATUS HOMO EX UTERO, BREVIORI TEMPORE VIVENS,
UT FLOS EGREDITUR, SED VELUT UMBRA FUGIT. Job xiv. 1, 2.

Man born of woman, living for a very short time,
Cometh forth like a flower, but fleeth as a shadow.

This is on a square-shaped dial, which is traced on the wall of a church, near Menaggio, that stands high above the Lake of Como on the western shore, and overlooks the houses beyond Menaggio and their white reflections in the water. If there is a date it was not observed. The dial is probably not old, and is placed close to the corner of the church wall, at an angle with the west, or rather south-west front, where there is a large porch supported on light pillars, and shaded by cypress trees.

174.

NESCIT OCCASUM LUMEN ECCLESIÆ.

The light of the Church knows no setting.

At Standish Vicarage, Gloucestershire. "I am with you always, even unto the end of the world."

175.

NIHIL VOLENTIBUS ARDUUM.

Nothing is difficult to the willing.

Is on a dial at Fyning House, in Sussex, which was erected in the reign of George II.

176.

NIL NI SIT SOL MI.

I am nothing unless I have the sun.

At Alzo, on the Lake of Orta, in North Italy.

177.

NIL NISI CÆLESTI RADIO.

Nothing without a ray from heaven.

Applicable alike to the dial, the church and the services, this motto is over the south door of the church of St. Mary the Virgin, at Lower Heyford, in Oxfordshire, where there has been a church from before the Conquest.

178.

N o light unthinking fondness, such as oft
E nshrines in pomp th' unworthiest of their line,
P rompted the tender thought, which here found words
T o tell of him we valued; one whose form
U nder this turf is mingled with the dust,
N o more to *live*; but whose recorded name,
E ndear'd to all, reminds us how to *love*.

N ear to this time-recording pillar's base
E ntomb'd, and, as became his merits, mourn'd—
P oor Neppy lies! the generous and the fond—
T he brave and vigilant—in whose nature shone,
U nited, all the virtues of his race:
N or grudged be this memorial, if its truth
E nforce the charge, "Be faithful unto death."
    Obiit Sep. 9, 1839, anno ætatis decimo.

In the garden of the Vicarage House at Borden, near Sittingbourne, Kent, there is a pedestal, surmounted by a sun-dial, which bears on its eastern and western sides two tablets inscribed with these acrostic epitaphs to the memory of a favourite Newfoundland watchdog, called "Neptune," by his sorrowing owner. These lines recall Lord Byron's "Inscription on the monument of a Newfoundland dog," dated "Newstead Abbey, Oct. 30, 1808."

179.

NON EXIGUUM TEMPUS HABEMUS, SED MULTUM PERDIMUS.

We have not a little time, but we waste much of it.

At Hatherley, in Gloucestershire. Sir W. Scott says otherwise:—

> "Redeem mine hours—the space is brief—
> While in my glass the sand grains shiver;
> And measureless thy joy or grief,
> When time and thou shall part for ever."

180.

NON NISI CÆLESTI RADIO.

Not except by a ray from heaven (do I tell the time).

On the church porch at Haydon Bridge, Northumberland. The dial is square, and the motto is above: the words being divided by a full-faced sun, which emits rays all round. They bear an obvious moral signification.

181.

NON REDIBO.

I shall not return.

CHARLES GREENWOOD, FECIT, 1790.

On a house in Westgate, Grantham.

182.

NON REGO NISI REGAR.

I do not rule unless I be ruled.

On the Crown Inn at Uppingham. The dial is square—black and gilt—and the motto acknowledges submission to the sun. It also illustrates the profound truth that—as à Kempis expresses it—" No man ruleth safely, but he that is willing to be ruled."

183.

NON SINE LUMINE.

Not without light (do I tell the time).

On the wall facing Leadenhall Street, London, of St. Catherine Cree Church, is a bold dial with this motto. The church seems to have been restored.

184.

NON TARDUM OPPERIOR.

I tarry not for the slow.

There is a stone figure of Time, bearded and with wings, on the terrace at Duncombe Park, Yorkshire, the seat of Earl Feversham, which is represented as about to carry away a vase-shaped pedestal, on the top of which is a dial thus inscribed. The figure, which is boldly carved, was the work of a local artist, the Helmsley Mason, about the year 1750, when we believe that the terrace was made.

185.

NONE BUT A VILLAIN WILL DEFACE ME.

Is to be seen at Kidderminster.

186.

NOS JOURS PASSENT COMME L'OMBRE.

Our days pass by like the shadow.

On a house in a street of Antibes, in the department of the Alpes Maritimes, is a dial painted on the wall, with the above motto. It is so close under the roof that the tiles overshadow it.

187.

NOSCE TEIPSUM, 1740, T. S.

Know thyself.

This is on Whitley Hall, an old Elizabethan house, in the parish of Ecclesfield, belonging to Thomas Shirecliffe in 1740: also on a house which stands near the bridge at Lewes; and on the cross-dial at Ellerslie, near Chichester.

188.

NOUS AVONS BESOIN DE PEU, ET POUR PEU DE TEMPS.

We need but little, and for a little time.

This sentiment is versified by Goldsmith in "The Hermit":—

"Man wants but little here below,
Nor wants that little long."

The motto is on a house near Aigle, in Canton de Vaud, Switzerland.

189.

NOW IS THE ACCEPTED TIME.

Is engraven over a stone dial, which rests on the top of a low wall, against an overshadowing tree, with water below. It is at Danby Mill, near Leyburn, Yorkshire, and strikes you when encountering it in so pretty and secluded a spot.

190.

NOW IS YESTERDAY'S TO-MORROW.

This quaint motto is on the porch of East Leake Church, in Nottinghamshire.

> "To-morrow, and to-morrow, and to-morrow,
> Creeps in this petty pace from day to day,
> To the last syllable of recorded time;
> And all our yesterdays have lighted fools
> The way to dusty death."
>
> *Macbeth*, Act v. Scene 5.

191.

NOW OR NEVER. 1514.

Is engraven on a vertical dial fixed on the top of a buttress of Monk Fryston Church, Yorkshire. It has also been deciphered on a square-faced dial, with four gnomons, and surmounted by a ball, which rests on a tall elegant stone shaft in Bolton Percy Churchyard, Yorkshire. It is on the south facing, and on the north side there is a faint trace of a former inscription, now wholly illegible from time and weather. The rector says, a tradition exists that the effaced words were "rationibus suis computandis," which may be supposed to be an exhortation to sum up your accounts. On Bolton Percy Rectory is an uninscribed dial, bearing the date 1698.

192.

NOW OR WHEN.

On a plain black and white dial, which is erected on the north-west tower of Beverley Minster.

193.

NULLA DIES SINE LINEA.

No day without its mark.

On the cross-dial at Ellerslie, near Chichester.

194.

NULLA FUIT CUJUS NON MEMINISSE VOLO.

There has been no hour which I do not wish to remember.

This is on a dial which is painted on the wall of a house at Fréjus, in the south of France, close beside a window of the third étage.

195.

NULLA FIAT CUJUS NON MEMINISSE JUVET.

May no hour occur which it will not delight me to remember.

At Bruges.

196.

NULLA HORA SINE LINEA.

No hour without its mark.

There are four dials in the cloisters of the cathedral at Chambery, all of large size; one on each side of the quadrangle. They are, as usual, traced on the plaster; but two of them are much broken and defaced. From their ancient appearance they have evidently borne many years of Alpine storms. The motto here quoted belongs to the dial on the south-east cloister. No date is on any of the dials; but old as they are, they must be considerably more modern than the adjoining cathedral, which was finished in 1430, in the reign of Amadeus III. (afterwards the Anti-Pope Felix V.), four years before his abdication and retirement to Ripaille, on the Lake of Geneva.

197.

NUNQUAM AURI,
SED OCULO
SÆPE GRATA. 1742.

Never acceptable to the ear, but often to the eye.

This motto may be read on a south-west declining dial, which is on the wall of the courtyard of the Mairie at Perpignan. It stands betwixt windows, some distance below the overhanging Spanish roof, whose border of greenish glazed tiles rests here and there on carved wooden owl-like figures, which project like gargoyles from the wall. There are two or three dials in this court over the low marble arcades, but only one bears an inscription. The building itself forms a part of the old Loge—from the Spanish "Lonja," or "Exchange of the Merchants." The façade, with its pointed arches, "exhibiting flamboyant ornaments, foliage, and tracery," dates from the fifteenth century. The carving is a good deal injured; the arches are now filled with glass, and that portion of the building is used as a café. It is one of the most remarkable structures in the old capital of Roussillon. The dial, however, is not of Spanish construction, as Perpignan came into the possession of the French in 1650.

198.

NUNC EX PRÆTERITO DISCAS.

Now you may learn from the past.

Is on Warrington School.

199.

O BEATA SOLITUDO, O SOLA BEATITUDO;
MIHI OPIDUM CARCER EST,
ET SOLITUDO
PARADISUS.
W ✠ H
DEUS NOBISCUM,
ET CORONA MANUUM OPUS NOSTRUM.
1663.
VIVAT CAROLUS SECUNDUS.

O blessed solitariness—O solitary blessedness: The town to me is a prison, and solitude my Paradise. O God, be with us, and crown the work of our hands. Long live Charles II.

The dial which is thus inscribed is circular, but has square ends below. The hours are marked round the lower half of the circle; and the lines of motto are above, and bend with the shape of the dial; the word "Paradisus" in the centre being straight, and from it the gnomon springs. It is formed of a single stone let into a plastered gable of a house fronting a garden in Priestgate, Peterborough, and belonging to Mr. G. Wyman. It was once held by a family named Hake, which may account for the initials W. H. There is a crown on the lower part of the dial face, which was gilt, but is now so much worn that only the general outline can be traced. The date and good wishes for the King are on the squared portion below. The obliging correspondent who favoured us with this interesting inscription, suggests that a very pretty story might be manufactured from it—" Of a poor old cavalier broken down in fortune and health; perhaps the object of persecution; perhaps the last of his family left after the civil wars, retiring to spend his last years in this secluded spot." Let us hope he may have known some bright hours in his loyalty, and perhaps joined in a glee with Roger Wildrake;—

> " Bring the bowl which you boast,
>   Fill it up to the brim;
> 'Tis to him we love most,
>   And to all who love him.

SUN-DIALS.

> Brave gallants, stand up,
>   And avaunt, ye base carles!
> Were there death in the cup,
>   Here's a health to King Charles!"

200.

O COONU CRE-CHA GIARE
AS TA MY HRRA.

O help my heart very shortly, and in my time.

"The small and great are there; and the servant is free from his master."—*Job* iii. 19.

UT HORA SIC VITA; DUM SPECTAS FUGIO.

As the hour, so is life; whilst you look I fly.

These mottoes, the first of which is in the Manx language, are on a sun-dial which stands at the gateway of St. Patrick's Church, in the village of Patrick, near Holm Peel, Isle of Man. The church was built by good Bishop Wilson in 1716.

201.

O TU, QUI BINAM UNO GNOMONE CONSPICIS HORAM,
HEU! MISER, IGNORAS QUA MORITURUS ERIS. 1822.

O thou who perceivest the double hour from one gnomon, alas! wretched one, art ignorant of the hour in which thou shalt die.

The "hours twain" referred to in this motto indicate the arrangement of the figures on the dial, which, besides telling the hour of the day in the usual manner, do also show the time by the Italian mode of reckoning: that is, as the hours are counted from sunset to sunset, going through the whole course of the twenty-four numbers. The lines of these additional hours, from xii to xxiv, which are traced upon the dial plane, declare the time by the shadow of a point in the stile, as it falls upon them. The dial itself is a large and wide one, the figures being represented in rolling clouds; and it is painted on a house wall that faces the sea at an opening of the main street of Cogoletto, a fishing village about eighteen miles west of Genoa, and a reputed birthplace of Christopher Columbus. The room in which, it is said, he was born, is still shown, and there are many inscriptions testifying to the tradition on the outside of the house, which were placed there by a member of the family in 1650. The tradition is firmly held by the inhabitants; and Washington Irving, who disputes the claims of all other places, save Genoa, nevertheless admits that there is some evidence in favour of Cogoletto. Admiral Colombo, with whom the great discoverer first sailed, was a native of this place; and the portrait of Columbus has been preserved here by his descendants. Tennyson seems to yield to this local claim:

> "How young Columbus seem'd to rove,
> Yet present in his native grove,
>   Now watching high on mountain cornice,
> And steering, now, from a purple cove,
> Now pacing mute by ocean's rim;
> Till in a narrow street and dim,
>   I stay'd the wheels at Cogoletto,
> And drank, and loyally drank to him."

202.

O WRETCHED MAN
REMEMBER THOU
MUST
DIE.
SENCE ALL THINGS
PASSE AND NOTHINGE
CERTAINE
BE.

This is one of the mottoes engraved on two sides of the square top to the solid pillar dial at Brougham Hall. On one of these sides is "Ut hora sic vita" with a skull carved below; and on the other side is

"Tempus ut umbra præterit," with an hourglass beneath. The two remaining sides are occupied with the date 1660: the initials $_{TE}^{B}$ (Thomas and Elizabeth Brougham) on one face, and on the other an armorial shield, which bears a fret, the coat of either Fleming or Hudlestone.

203.

OF SHADE AND SUNSHINE FOR EACH HOUR
SEE HERE A MEASURE MADE:
THEN WONDER NOT IF LIFE CONSIST
OF SUNSHINE AND OF SHADE.

On a pleasant villa house at Wadsley, near Sheffield, called "Dial House," is a stone dial let into the wall, which bears the foregoing inscription.

204.

OH QE LE TEMP PASSE VITE!

Oh, how quickly time passes.

This is badly spelt, and appears over the door of a small house, which stands in a garden a little off the road between Cannes and Grasse. The dial is round, facing south, with a green background. It represents the sun's full face, broad and smiling, with his hair dressed after the fashion of a king in a pack of cards. He holds the gnomon like a pipe in the corner of his mouth, and seems to be regretting the swift passage of a jolly life.

205.

OMNES TIME PROPTER UNAM.

Fear every hour because of one (the last).

At Visp in Switzerland.

206.

ON THIS MOMENT HANGS ETERNITY.

The chapel of Alfrick, in Worcestershire, has a dial with this motto upon it.

207.

ONCE AT A POTENT LEADER'S VOICE I STAY'D,
ONCE I WENT BACK WHEN A GOOD MONARCH PRAY'D;
MORTALS: HOWE'ER WE GRIEVE, HOWE'ER DEPLORE,
THE FLYING SHADOW WILL RETURN NO MORE.

Taken from Cyrus Redding's "Fifty years' Personal Recollections."

208.

OPPOSTO DI ME,
PENSI DI TE.

Opposite to me, think of thyself.

At the chateau of the Count Pinsuti in Piedmont.

209.

ORA ET LABORA.

Pray and work.

On the church at Northallerton, Yorkshire. The dial is very plain in form.

210.

ORA, NE TE RAPIAT HORA.
Pray that the hour snatch you not away.

This is inscribed high up on the tower of a large modern church in a small village of the Val Sesia, near Varallo, North Italy. The village has narrow shady streets, steep-roofed houses, and garden walls with vines straggling over them.

211.

ORDINATIONE TUA, REGE ET PROTEGE.
By thy ordination, rule and protect.

On Visp church in the Rhone Valley, Switzerland. The motto is on the dial face, which also contains the sun and a church painted on it.

212.

ORIENS EX ALTO VISITAVIT NOS:
MEMOR ESTO OCCASUS TUI.
The day star from on high hath visited us:
Be thou mindful of thine own setting.

The first line is on the east face, the second on the west face, of a dial at Round House Farm, Haverfield, Gloucestershire.

213.

ORIMUR, MORIMUR,
We rise up, we die.

On a gable at Packwood Hall, Gloucestershire. We are told that when this motto was last painted, the artist unfortunately put *mortimur* for "morimur." We cannot doubt that this is a true account of the position, for a sketch of the square-shaped dial, immediately under a small window in the angle of the gable, is before us, with the legend below. An obliging communication, however, from Bishop Hobhouse, informs us that the same words are on a clock-face at Packwood; the word "orimur" being over an increasing series of figures, and the word "morimur" over a decreasing series.

214.

OUR DAYS ON EARTH ARE AS A SHADOW:
SO SOON PASSETH IT AWAY, AND WE ARE GONE.

In the gardens fronting the house at Gale Syke, Wastwater, is a pillar-dial thus inscribed, that was erected about twenty years ago, and presented to the owner of the place, Stansfield Rawson, Esq., by one of his daughters. It was probably designed by Mr. Rawson's son-in-law, the Rev. Dr. Worsley, Master of Downing College, Cambridge, and is very tasteful.

215.

OUR DAYS PASS LIKE A SHADOW.

Is on the church at Whitby, Yorkshire.

216.

OUR LIFE'S A FLYING SHADOW, GOD'S THE POLE:
THE INDEX POINTING TO HIM IS OUR SOUL;
DEATH'S THE HORIZON WHEN OUR SUN IS SET,
WHICH WILL, THROUGH CHRIST, A RESURRECTION GET.

This may be read on a tablet under the porch dial at Ebberston Church, near Scarborough, dated 1843; also on Milton Church, Berkshire, 1859; and on Glasgow Cathedral, which has other inscriptions. The second and third lines are in some cases transposed.

217.

PARTE L' OMBRA COL SOL, COL SOL RITORNA:
MA L' UOM QUAL OMBRA FUGGE, E PIU NON TORNA.

The shadow departs with the sun, with the sun returns:
But man as the shadow flees, and returns no more.

At Sordevole, in Piedmont.

218.

PEREUNT ET IMPUTANTUR. $_{J.A.}^{T.}$ 1861.

They perish and are reckoned.

This is one of the Temple dial mottoes, in Temple Lane. Our drawing has the date 1818, T. A. H.; but this was made prior to the repainting in 1861, when the Prince of Wales visited the place. The same motto is over the south porch of Gloucester Cathedral, shaded by canopy work; also on a dial which is fixed against the battlements of All Souls' College, Oxford, betwixt two pinnacles, where it was placed by Sir Christopher Wren, who was a fellow of the College. It is also on Lincoln Cathedral, and at Bamborough, Northumberland; as well as at Kildwick Church, near Skipton, in Craven. It may be read on the curious clock at Exeter Cathedral. The words are taken from an epigram by Martial, the four last lines of which are as follow:

"Nunc vivit sibi neuter, heu! bonosque
Soles effugere atque abire sentit,
Qui nobis *pereunt et imputantur;*
Quisquam vivere cum sciat, moratur?"

Cowley translates these—

"Now to himself, alas! does neither live,
But sees good suns of which we are to give
A strict account, set and doth march away:
Knows a man how to live, and does he stay?"

The sentiment is remarkable from a heathen writer, and somewhat more

Christian, though often not more true, than that given to a lady who was being lionized at Oxford, and asked the meaning of the words—"They perish," said her waggish companion, "and are not thought of." *Pereunt, imputantur* is on the cross-dial at Elleslie, near Chichester. *Pereunt et imputantur* was on Rotherham Church, Yorkshire, but the dial motto is now, *Hora fugit: memento mori*: "The hour flies, remember you have to die."

219.

PETITO QUOD JUSTUM.

Seek what is just.

In Jamaica there is an old Spanish sun-dial placed on the parapet of the platform, before the main entrance to Great Pond House, parish of St. Anne, just in front of a pomegranate tree, which springs from the rock opposite the dial. The dial is inscribed as above.

220.

PORTATRICE A VOI DI BENE,
L' ORE SIAM DE' DI SERENI,
SI ANNOTTA, O TUONA, O PIOVE
NOI FUGGIAM IN GREMBO A GIOVE.

Bearers of wealth to you, the sons of men,
Are we, the sunlight hours of days serene;
If night, or rain, or thunder blur the sky,
Into our Father's bosom back we fly.

So the late Dean Alford translated the motto which he found at Vignale, on the Riviera.

221.

POST EST OCCASIO CALVA.

Opportunity is bald behind.

"Take Time by the forelock," says the adage. The whole line is—

"Fronte capillatâ, post est occasio calva," as already stated. Lord Bacon, in his Essay "Of Delays," thus renders the verse—"Occasion turneth a bald noddle, after she hath presented her locks in front, and no hold taken." The above motto is on both Yaxley, Hants, and Horton, Dorsetshire, churches.

222.

POST TENEBRAS LUX.

After darkness light.

A modern dial, near the corner of a house, with a high garden wall at Varenna, on the Lake of Como, bears this inscription. The house stands at the entrance of Varenna, which is a bright little town by the water side, with a ruined castle on the crag above; and it is sacred to Manzoni, who has given an accurate description of the scenery in his account of Don Rodrigo's Castle—*see* "I Promessi Sposi," chap. v.

223.

POST TENEBRAS SPERO LUCEM.

After darkness I hope for light.

Is recorded by Mr. H. Hopley, but no locality is assigned.

224.

PRÆSTANT ÆTERNA CADUCIS.

The things eternal excel the perishable.

Noted in North Italy.

225.

PRÆTEREUNT. IMPUTANTUR.

They pass by. They are reckoned.

There are two sun-dials at Farnham Castle, on the walls of the entrance tower. They had formerly the inscription "Eheu, fugaces labuntur anni;" but have now the more befitting words "Prætereunt," on the one, and "Imputantur" on the other.

226.

PRÆTERITUM NIHIL,
PRÆSENS INSTABILE,
FUTURUM INCERTUM.

The past is nothing : the present unstable : the future uncertain.

Once in the pleasure ground of Knole Park, Sevenoaks, stood a costly, but rather inelegant, white marble pillar dial, the gnomon of which is supported by an earl's coronet. It has been removed to the garden of a neighbouring farm house, and bears the foregoing motto. Knole was the property of the Duke of Dorset, whose co-heiresses were his two sisters, the elder of whom married the Earl of Plymouth, and inherited the estate. She afterwards became the Countess Amherst. In default of issue, the estate passed to the younger sister, the wife of Earl Delawarr, who was created Baroness Buckhurst in her own right, with remainder to her second son, who is now in possession of Knole.

227.

PREPARE TO MEET THY GOD. LAT. 53 DEG. 26 MIN.
TEMPUS FUGIT UT UMBRA.

Time flies as a shadow.

GLORIA DEI.

The glory of God.

These mottoes are on the porch dial of Bradfield Church, in the parish of Ecclesfield, Yorkshire. The church is a fine ancient structure, and is nobly placed, overlooking the moors and the valley, through which the flood, caused by the bursting of the Dale Dyke Reservoir, poured down into Sheffield in the night of March 11, 1864, when 250 persons were drowned.

228.

PROPERAT HORA MORTIS:
ULTIMA CUIVIS EXPECTANDA DIES.
LUX UMBRA DEI.
DUM SPECTAS FUGIO.
TENERE NON POTES, POTES NON PERDERE DIEM.

The hour of death hastens on: the last day is to be looked for by each one. Light is the shadow of God. Whilst you look, I fly. You cannot hold, you cannot destroy a day.

This motto, preceded by the name, "Johannes Watkins, 1695," was formerly at East Harptree, and is now in the possession of the Rev. H. Hooper, at Ripley, near Guildford.

229.

PULVIS ET UMBRA SUMUS.

We are dust and shadow.

In Leyland churchyard, Lancashire.

230.

QUA REDIT NESCITIS HORAM.

You know not the hour in which he returns.

On an old gable in Lincoln's Inn there is a western dial which is thus inscribed. It was restored in 1794, when the great William Pitt was treasurer; and again renovated in 1848. The aspect allows only the hours from noon to evening to be recorded.

231.

QUALIS VITA FINIS ITA. 1809. W. DEACON.

As your life is, so shall your end be.

At Great Smeaton; and in smaller characters—

OMNIS SPIRITUS LAUDET DOMINUM.

Let everything that hath breath praise the Lord.

Together with—

LEX DEI, LUX DIEI.

The law of God is the light of day.

See No. 143.

232.

QUAM CITO JUCUNDI PRÆTERIERE DIES.

How soon the pleasant days have passed away.

No place recorded.

233.

QUANDO DI NUBI AL SOL SGOMBRA É LA VIA, ALLO STANCO VISITANTE ADOTT' É L' ORA CHE LO CHIAMA AL RISTORO E ALL' ALLEGRIA.

When the path of the sun is free from clouds,
To the weary traveller is brought round the hour
Which calls him to refreshment and mirth.

Beyond Varenna the road to Colico winds along the shores of the Lake of Como, and passes a little roadside *osteria*, over the door of which is a rough sun-dial with the above motto. It serves as a sign to the inn, as well as to indicate the time. The *osteria* stands almost at the entrance of the tunnel through the cliff which borders the lake. It may have been built for the use of the workmen employed in the road-making, of which these tunnels were the longest and most difficult part. The work was finished about 1826, and thereby the great Stelvio road was completed.

234.

QUI LUCEM DE TENEBRIS LUCET IN CORDE.

He who sends light from darkness shines in our heart.

"God who commanded the light to shine out of darkness hath shined in our hearts," 2 Cor. iv. 6, will have suggested this motto, which is on the old Grammar School at Wellingborough, near the church; and there is the further inscription:—

ἩΜΕΡΑΙ ΩΣΕΙ ΣΚΙΑ.

Our days are as a shadow.

A sun with rays occupies the upper part, and the Latin line is written round it.

235.

QUID CELERIUS TEMPORE?

What is swifter than time?

One of the mottoes on the cross-dial at Elleslie, near Chichester.

236.

QUOD PETIS UMBRA EST.

What you seek is a shadow.

Is at Hebden Bridge, near Halifax, Yorkshire.

237.

REDIBO, TU NUNQUAM.

I shall return, you never.

At Erith, in Kent. It is also one of the Rev. W. L. Bowles' suggested mottoes, and he applies it to the "setting sun," giving the following paraphrase:—

"Haste traveller, the sun is sinking now:
He shall return again, but never thou."

See Nos. 65 and 300.

238.

REMEMBER. 1803.

At West Ham.

239.

RESPICITE, NON MIHI SOLI LABORAVI. 1593.

Mark, not for myself alone have I laboured.

On the triangular Lodge at Rushton, Northamptonshire, which was built by Sir Thomas Tresham, and is an architectural curiosity. The building is in fine preservation, and has on each side three gables, which severally terminate in a pinnacle, and on the centre gable of each side there is a sun-dial with an inscription. On one is the word "Respicite;" on another, "Non mihi;" and "Soli laboravi" on the third. The plan of the building is symbolic of the Trinity, which is also expressed in the trefoil that forms part of the family crest.

Sir Thomas Tresham, who founded this lodge on his estate, was knighted by Queen Elizabeth at Kenilworth Castle; but being a firm adherent to the Roman Church, like his ancestors before him, he suffered a long imprisonment in the Castle of Wisbech for recusancy. Indeed, for this offence he was three times in custody; his last commitment being on the 31st Dec. 1596, from which he was discharged by warrant on the 8th Dec. 1597. He was a skilful architect, and built the market-house at Rothwell. "Having many daughters," says Fuller, "and being a great housekeeper, he matched most of them into honourable, the rest of them into worshipful and wealthy families." They were six in number. The following extract from a letter, written by Sir Thomas Tresham, about 1584, is curious:—
"If it be demanded why I labour so much in the Trinity and Passion of Christ, to depaint in this chamber, this is the principal instance thereof; That at my last being hither committed, and I usually having my servants here allowed me, to read nightly an hour to me after supper, it fortuned that Fulcis, my then servant, reading in the 'Christian Resolution,' in the treatise of 'Proof that there is a God, &c.' there was upon a wainscoat table at that instant three loud knocks (as if it had been with an iron hammer) given, to the great amazing of me and my two servants, Fulcis and Nilkton." This story remains to show that there is nothing new under the sun—not even "table rapping." The triangular Lodge is rich in pious emblems and inscriptions—a noble monument of Sir Thomas's zeal for Trinitarian doctrine. His family was ancient and influential by wealth and character.

240.

RES SACRAS CLERI, THEMIDIS, MARTISQUE LABORES,
ET PATRIOS CŒTUS, LUMEN ET UMBRA REGIT.

My light and shade regulate the sacred affairs of the clergy, the labours of the law and of war, and the national assemblies.

The Cathedral of Fréjus in the Department du Var, France, is a Romanesque building of the eleventh and twelfth century. Over one of the doors, with ivy growing up the side, is a wooden dial, painted blue with gilt lettering, and thus inscribed.

241.

RITORNA IL SOL DALL' OMBRA
SPARITA:
MA NON RITORNO, NO, L'ETÀ
FINITA.

The sun returns which has been displaced by the shadow: but there is no return, none, of Time gone by.

This is on a house, 22, Via Gregoriana, in Rome. Also on the wall of the Douane at Isella is the same motto, but varied in the second line, which runs thus:—

MA NON RITORNA PIU L'ETA FUGITA.

But there is no return of Time which has flown.

The writing, which is very much defaced, may be just read between arches that run along the front of the building. This stands close upon the great Simplon road, and is the first or last halting-place on the Italian frontier. Standing beside it, you may look directly towards the gorge of Gondo, and at the great white peak which glitters high above in the sunshine, backed by a deep blue sky. No date was noticed; but the road itself was finished in 1805, having been begun by Napoleon's order after the battle of Marengo in 1801.

SUN-DIALS.

242.

SCIS HORAS, NESCIS HORAM.

You know the hours, you know not the hour.

On the convent of Cimiès, near Nice.

243.

SE IL MOTO TALOR NON SEGNO L' ORE,
DELLA NATURA SOLO È COLPA IL GIOCO;
SE SI SPIEGA LA CENSURA IL SUO FURORE,
INVIDIA E NON RAGION VI PUÒ DAR LOCO.

L' ETERNO FACITOR CON GIUSTO PESO
L' OPRE SUE COMPARSE CON MISURA;
SE IN CIÒ CREDE TALUN D' ESSERNE LESO,
STOLTO! L' OPRE DIVINE ALLOR CENSURA.

SE IL SOL RESPLENDE IN PIENO GIORNO,
SE IL MONTE OPPOSTO NON TOGLIE I RAGGI,
VERGATE ORE AVRAI A TE D' INTORNO,
QUETI E MUTI FARÒ I MENO SAGGI.

If the movement (of the shadow) does not point out the hours, the freaks of Nature are alone in fault. If the anger of the critic is thereby aroused, envy and not reason is the cause.

The eternal Creator with just weight[1] and with (perfect) measure has shown forth his works. If any one considers himself wronged by them; Fool! they are the Divine works which he censures.

If the sun shines in full brilliancy—if the mountain in front does not shut out

---

[1] This idea is evidently that of Genesis i. 25, and the expressions seem to be borrowed from Job xxviii. 25. "Quando Egli dava il peso al vento, e pesava l'acque a certa misura."

the rays, thou wilt have the hours written around thee, and even the most foolish will be silenced and convinced.

These inscriptions are the mottoes on three separate dials, which are placed on the sides of the tower of the Campanile at Trafiume, near Canobbio, on the Lago Maggiore. Near the top of the tower is the date thus given:— " Anno Domini MDCCCVIII Ristaurato nel 1808. Capo Maestro Andrea de Bernardi." Each verse is written below its respective dial. We may assume there were dials before the year named.

244.

SE IL SOL BENIGNO MI CONCEDE IL RAGGIO,
L' ORA TI MOSTRA, E IL CIEL TI DIA BUON VIAGGIO.

If the kind sun grants me the ray,
It shows thee the hour: and may Heaven give thee a good journey.

Was read somewhere on the route between Florence and Bologna.

245.

SE ME MIRAS, ME MIRAN.

If thou lookest at me, they look at me.

In Spain.

246.

SEE THE LITTLE DAYSTAR MOVING—
LIFE AND TIME ARE WORTH IMPROVING,
  SEIZE THE MOMENTS WHILE THEY STAY:
SEIZE AND USE THEM,
LEST YOU LOSE THEM,
  AND LAMENT THE WASTED DAY.

Mr. Howard Hopley has recorded this motto, without naming the locality; and he says, " The little daystar was a spot of light falling through a hole in the pointer, to indicate the hour."

247.

SEGNO SOLO LE ORE SERENE.

I only mark the bright hours.

This motto is on a modern dial on the wall of the back of Villa Novello, at Genoa. A bank separates the house from the sea, which it faces, and is luxuriantly covered with aloes, prickly pear, and other plants so familiar along the Riviera. The strong north wind sweeps down, and prevents the same careful cultivation of this spot which prevails in the rest of the garden; but its wild luxuriance is very charming.

248.

SEMITAM, PER QUAM NON REVERTAR, AMBULO.
JOB, XVI.

I tread a path by which I shall not return.

The reference is evidently to the words " When a few years are come, then I shall go the way whence I shall not return," Job xvi. 22. The dial is modern, and without date. It is painted yellow, bordered with grey. It is placed south declining west, on the wall of the church at Lavagna on the Riviera di Levante. The gnomon comes from the mouth of the sun, as if it were the tongue with which he was speaking. Lavagna lies between Chiavari and Sestri Levante. It formerly belonged, together with the greater part of the east coast, to the Fieschi of Genoa, who bore the title of Counts of Lavagna—notably so to that Gian Luigi Fieschi, who was the author of the famous conspiracy against Andrea Doria. The dial is beside the great door of the church, to which a flight of marble steps leads up, and which faces the sea.

249.

SEMPRE A VOI SEGNI ORE TRANQUILLE IL SOLE,
QUASI RAGGIO DI LEI CHE QUI SI COLE.

*May the sun always mark tranquil hours to you,
As it were a beam from Her who is honoured here.*

At Villa Mylius, Genoa, there is a plaster cast of the Madonna and Child, and immediately above the figures a dial is painted on the wall with the above motto.

250.

SENESCIS ASPICIENDO.

You grow older whilst you look.

On a dial at Versailles.

251.

SENZA L' OMBRA NON DILETTO,
E PUR L' OMBRA È MIO DIFETTO.

*Without a shadow I do not please;
Nevertheless a shadow is my defect.*

At Strevi, Monferrato.

252.

SENZA PARLAR IO SONO INTESO,
SENZA RUMOR L' ORE PALESO.

*Without speaking I am understood.
Without noise I reveal the hours.*

At Sordevole in Piedmont, and at Bordighera. Dean Alford notes and translates this motto:—

SUN-DIALS.

> "I speak not, yet all understand me well;
> I make no sound, and yet the hours I tell."

A gentleman walking from Como to Monte Generoso observed at Balerna a slightly different version.

SENZA PARLAR DA TUTTI SON INTESO,
SENZA FAR RUMOR L' ORA PALESO.

Without speaking I am understood by all:
Without making a noise I reveal the hour.

253.

SEPTEM SINE HORIS.
Seven without the hours.

The meaning of this bald inscription must be, that there are, in the longest days, seven hours (and a trifle over) in which the dial is useless. The motto is on a dial erected on a gable at Packwood Hall, in the county of Warwick. See No. 213.

254.

SET ME RIGHT AND USE ME WELL,
AND I Y^E TIME TO YOU WILL TELL.

Is engraved on an old pocket dial, which its owner thus describes: "It is a ring of brass, much like a miniature dog-collar; and has, moving in a groove in its circumference, a narrower ring with a boss, pierced by a small hole to admit a ray of light. The latter ring is made moveable to allow for the varying declination of the sun in the several months of the year, and the initials of these are marked in ascending and descending scale on the larger ring which bears the motto. The hours are lined and numbered in the opposite concavity."

255.

SHADOWS WE ARE, AND LIKE SHADOWS DEPART.

On a dial in Pump Court, Temple, which was restored in 1861.

256.

SI SOL DEFICIT, NEMO ME RESPICIT.
If the sun fails, no one regards me.

In the cloisters at Chambery. See Nos. 61, 77, and 196.

257.

SIC LABITUR ÆTAS. 1778.
GARGRAVE FECIT.
Thus life slips away.

At Middleham Church, Yorkshire: also at Darlington.

258.

SIC MEA VITA FUGIT.
So my life flies.

This is at Asti, an ancient town between Turin and Alessandria.

259.

SIC SUA CUIQUE DIES.
So is his day to each one of us.

PARAPHRASE.
"Thus every passing life is found
A passing shadow on the ground."

On a dial in a village between Lugano and Como.

260.

SIC TRANSIT GLORIA MUNDI.

So passes the glory of the world.

This may be read on Fountains Hall by Studley Park, near Ripon. The hall was built out of the ruins of the adjacent abbey by Stephen Proctor, one of the esquires to James I. The same motto is over the church porch at St. Just, Cornwall, where the dial is made of slate; and over it is a representation of an angel holding an hour-glass, with the sun half-risen. Below is the name of Nicholas Raleigh. The same inscription is on Louth Church, Lincolnshire, and on the Convent of Pomier, near Geneva, also on the Cross-dial at Elleslie, near Chichester.

261.

SIC UMBRÆ DECLINAVERUNT.

So have the shadows gone down.

On the high campanile of a church near Lugano.

262.

SIC VITA.

So is life.

On Threckingham Church, Lincolnshire.

263.

SIC VITA FUGIT.

So life flies.

At Sestri Levante on the Riviera.

264.

SIC VITA TRANSIT.

So passes life.

The tourist who penetrates to the sequestered spot, may read this dial motto on the old house of Compton Wynyates near Edgehill in Warwickshire, which belongs to the Marquis of Northampton, and is sometimes called "Compton in the Hole" from its position; as it stands in a deep hollow, surrounded by hills and woods, and seemingly shut in to perpetual loneliness. It is a grand old hall, and was built by Sir William Compton (temp. Henry VIII.), who is said to have brought the curious chimneys from the Castle of Fulbrook which he demolished. He stood in the favour of his king, and may be said to have founded the Compton family, as noble. His grandson was created Earl of Northampton by James I., and was father of the "loyal Earl" who followed Charles I., and grandfather of Compton, Bishop of London, who opposed James II. The old house suffered much in the civil wars, and is now dismantled; being only visited at intervals for a few days by its owners. It is built round a court, and surrounded by a moat. The roof and ceilings are in good repair. It contains a small chapel for secret celebration of the mass with private staircases. The dial hangs on the east side of the house, overlooking what was formerly the pleasaunce. Mr. Howitt writes: "When about a furlong from the house, I turned and saw that it was already hidden in its deep comb, and shrouded by its wooded hills; and I was deeply sensible of the utter loneliness and silence of the scene. I never on the moors of Scotland or Cornwall felt such a brooding sense of an intense solitude."

265.

SIC VITÆ CERTA RATIO:
TEMPUS FUGIT, MORS VENIT. 1747.

Such is the sure reckoning of life :
Time flies : death approaches.

At Brough, fixed on a tombstone-shaped stone.

266.

SICUT UMBRA.

As a shadow.

A small square dial, fixed at an angle to the wall over the arch of the doorway of the south porch of Maker Church near Devonport, contains this motto.

267.

SINE FEBO . . . EST NIHIL.

Without the sun (Phœbus) . . . it is nothing.

At Castel del Pazzo, near Rome, the inscription is partially defaced: three letters are apparently gone.

268.

SINE SOLE NIHIL.

Nothing without the Sun.

At Puisseaux in France.

269.

SINE SOLE SILEO.

I am silent without the Sun.

On the chapel of St. Philippe, Nice; also at Pino in Piedmont, and at Alghero in Sardinia.

270.

SIT PATRIÆ AUREA QUAVIS.

May there be for our country some golden (hour).

There is a dial carved in stone above the façade of the Maison du Roi, or Broodhuys, at Brussels, which bears this motto. It is on the fine old building, opposite the Hotel de Ville, in the square where the executions of Counts Egmont and Hoorn took place. Both Prescott and Motley write as if the present building was the same in which these two noblemen were confined before their execution. Prescott says, " The prisoners were at once conducted to the Broodhuys, or 'Breadhouse,' usually known as the Maison du Roi—that venerable pile in the market-place at Brussels still visited by every traveller on account of its curious architecture, and yet more as the last resting-place of the Flemish lords. Here they were lodged in separate rooms, small dark, and uncomfortable, and scantily provided with furniture."—*Hist. of Philip II.* b. iii. c. 4.

Motley also writes, " The Count (Egmont) was confined in a chamber on the second story of the Broodhuis, the mansion of the crossbowmen's guild, in that corner of the building which rests on a narrow street running back from the great square."—*Rise of the Dutch Republic*, vol. I. p. 173.

Egmont and Hoorn were executed in 1568. In a scarce little book called " Les Délices des Pays Bas," published in 1769 at Liege, there is this stated: " La maison du Roi, qu'on Broodhys fut batie en 1618 par l'ordre de l'Archiduc Albert et d'Isabelle; mais les plus beaux ornaments ont été gâtés au bombardment." It seems probable that this account is the correct one, because of the inscription which runs along the whole façade, one line below each story, " A peste, fame et bello, libera nos, Maria Pacis—ac votum pacis publicum—Elizabeth consecravit." This Elizabeth is evidently Isabella, the daughter of Philip II., wife of Archduke Albert, to whom Philip granted the sovereignty of the Netherlands in 1598.

271.

SIT SINE LITE DIES.

Let the day be without strife.

On Darlington church. The dial is placed against a built-up window; the face is black, the lines and lettering gold.

272.

SO FLIES LIFE.

On an old house at Southall, Middlesex.

273.

SO FLYS LIFE AWAY. 1738.

On the church at South Stoneham, Hampshire: "Jo. Sharpe, Ro. Houghton, Churchwardens."

274.

SO ROLLS THE SUN, SO WEARS THE DAY,
AND MEASURES OUT LIFE'S PAINFUL WAY:
THROUGH SHIFTING SCENES OF SHADE AND LIGHT,
TO ENDLESS DAY OR ENDLESS NIGHT.

FOR THE LADY ABNEY AT NEWINGTON 1735.

These lines were written by Dr. Watts as the motto on a handsome solid pillar-dial which formerly stood in the garden of Lady Abney at Stoke Newington, Dr. Watts being resident there as tutor in the family of Sir J. Hartopp. Sir Thomas Abney was Lord Mayor of London in 1700, and died in 1722. The dial has been removed to Edmond Castle, near Carlisle, the residence of J. H. Graham, Esq. Mr. H. Hopley has noted a different version of the lines, without recording any locality:—

"So glide the hours, so wears the day,
These moments measure life away,
With all its trains of Hope and Fear;
Till shifting scenes of Shade and Light
Rise to Eternal Day, or sink in endless Night."

275.

SOL GLORIA MUNDI.

The sun the glory of the world.

On a house in Whitehorse Yard, Wellingborough. It is a square wooden dial, without date.

276.

SOL ME, VOS UMBRA.

The sun (guides) me, the shadow you.

On a farmhouse at Coldthorpe, Gloucestershire.

277.

SOL TEM[P]O DI SATURNO IL DENTE EDACE
E DEL PALLONE IL GIOCATOR FALLACE. 1826.

At Chieri, in Piedmont. Two or three Italians have tried to make sense of this obscure motto, and have failed. The first allusion to the mythological legend of Saturn devouring his children will be recognized; and the accompanying simile can only be explained by a reference to the favourite Italian game of *pallone*. This game somewhat resembles *tennis*, and still remains a living representative of the old Roman game of *pila*. The manner of playing it has been thus described by Mr. Story ("Roba di Roma," vol. i.): "It is played between two sides, each numbering from five to eight persons. Each of the players is armed with a *bracciale*, or gauntlet of wood, covering the hand and extending nearly up to the elbow, with which a heavy ball is beaten backwards and forwards, high into the air, from one side to another. The object of the game is to keep the ball in constant flight, and whoever suffers it to fall dead within the bounds loses. The game is played on an oblong figure, marked out on the ground, or designated by the wall around the sunken platform on which it is played, and across the centre is a transverse line dividing the two sides; and as the ball falls here and there, now flying high in the air, and caught at once by the *bracciale* before touching the ground, now glancing back from the wall which generally forms one side of the lists, the players rush eagerly to hit it, calling loudly to each other, and often displaying great agility, skill, and strength." Allusions to the game of *pallone* may be found in the works of the modern Italian poets. Leopardi and Aleardi have both made use of it as a subject for their verse.

The above motto was ultimately shown to Antonio Fraschio, a gondolier in the service of the National Bank at Venice, who is well known for his interpretation of Dante's *Divina Commedia*, and is exceedingly clever. He said at once that the word *tempo* should be *temo*, and then the meaning would be, "I only fear the devouring tooth of Saturn, and the inexpert player with the ball." That is, the gnomon fears alike Saturn's wet weather which corrodes iron, and the bad pallone player who may throw his ball against and break it.

278.

SOL TIBI SIGNA DABIT: SOLEM QUIS DICERE FALSUM AUDEAT?

The sun will give you the signs: who dares to say the sun is false?

From Virgil's first Georgic, line 463, and is the motto of a sun-dial on one of the terraces at Bramshill Park, Hampshire, the seat of the Rev. Sir William H. Cope, Bart. At the same place are three other dials, but only the family arms with date and initials are engraved on them. It was the motto of the old "Sun" newspaper.

279.

SOLE ORIENTE, FUGIUNT TENEBRÆ.

With the rising sun the shadows flee.

On a dial in a garden in the diocese of Connor.

Bishop Mant, in his Latin and English poem, "The Sun-dial of Armoy," writes thus:—

"Night flies before the orient morning,
 So speak the Dial's accents clear;
So better speaks the Prophet's warning
 To ears that hear.

"Night flies before the Sun ascending;
 The sun goes down, the shadow spreads—
Oh, come the day which, never ending,
 No night succeeds.

> "And see, a purer day-spring beaming,
> Unwonted light, nor moon nor sun;
> But Light itself, with glory streaming,
> God on His throne!"

280.

SOLES PEREUNT ET IMPUTANTUR.

Days (suns literally) depart and are reckoned.

Outside the Dean's kitchen, at Durham, is a square dial, gold-lettered, which bears this inscription.

281.

SOLI, SOLI, SOLI. 1756.

On a simple south dial, with the motto and date on a scroll above and the face of the sun made a central point to which the gnomon is fixed, is the word "Soli" thrice given. We must leave the reader to translate this to his own fancy. It occurs at Monthey, which is a small town in the Valley of the Rhone, in the Canton du Valais, at the entrance of the Val d'Illiez. The same may also be read at Bonneville.

282.

SOLIS ET ARTIS OPUS.

The work of the sun and art.

Is the motto on a house-dial at Grasse in the Department of the Alpes Maritimes. The dial is a large oblong, plain in pattern, and painted on the wall with the motto at the top. The same may also be seen at Milan on a dial, south declining west, which is modern and without date.

283.

SON FIGLIE DEL SOLE,
EPPURE SON OMBRE.

They (the hours) are daughters of the sun, and yet are shadows.

So we venture to print this motto, for as it stands in Dean Alford's book, there is clearly some mistake in the wording; and we have now, by the alteration of a single letter, "figli*e*" for "figli*a*," made what appears to us better sense of the whole. The Dean gives the following paraphrase:

"I the sun my father call,
Yet am shadow after all."

284.

SON POCHE LE ORE MIE, LE TUE SON MOLTE.

My hours are few, thine are many.

At St. Remo on the Riviera.

285.

SPECTATOR FASTIDIOSUS, SIBI MOLESTUS.

The spectator, who is too curious, only causes trouble to himself.

At Bywell Abbey, near Newcastle-on-Tyne. A translation is difficult, and the meaning is obscure.

286.

STEH' BEY UNS IN ALLER NOTH,
HIER IN LEBEN UND IN TOD.

Stand by us in all need, here in life and in death.

At Salzburg. The dial contains an elegant design of the Madonna holding the Divine Infant in her lap, who appears to be guiding the gnomon.

287.

SUB UMBRA QUIESCUNT,
SUB LUCE GAUDENT. 1770.

Under the shade they rest, (in death)
Under the light they rejoice.

On the "Hôtel des Invalides" at Paris.

288.

SUPREMA MULTIS HORA, FORSAN TIBI.

The last hour to many, perhaps to thee.

On the Riviera.

289.

TACITO PEDE LABORO.

I labour with a silent foot.

This dial motto is on the outside wall of the old palace of the princes of Masserano (La Marmora) at Masserano in the province of Novara in Italy. Lamartine expresses a similar idea:—

" L'ombre seule marque en silence
Sur le cadran rempli, les pas muets du temps."

290.

TEMPORA MUTANTUR, NOS ET MUTAMUR IN ILLIS.

The times are changing, and we are changing in them.

On a pillar-dial in the garden at Brockhampton Park, near Cheltenham. This well-known line is not classical, and is stated to have been a saying of Lotharius I. (*flor. cir.* 830 A.D.), the true reading being " Omnia mutantur," &c.

291.

TEMPORA TEMPORE TEMPERA.
Moderate the times in time.

On the church at Vian in Piedmont.

292.

TEMPORE NIMBOSO SECURI SISTITE GRADUM,
UT MIHI SIC VOBIS HORA QUIETIS ERIT.

In a stormy time stay your step, secure,
As to me, so to you, it will be an hour of rest.

This pretty and appropriate inscription is placed above a plain dial, south declining west, which is painted on the side wall of an Inn near one of the stations on the Mont Cenis railway line, just before reaching St. Michel, and nearly at the foot of the mountain. It will be seen that the word *gradum* will not scan where it stands.

293.

TEMPORIS MEMOR MEI, TIBI POSUI MONITOREM. CHRISTIAN.
1681. DE WHITEHOUSE.

Mindful of my own time, I have erected an admonisher for you.

The above inscription was kindly furnished by a correspondent, as being the motto on a dial in the churchyard of Kirk Michael, in the Isle of Man. He described the words as "much worn," and was not quite certain about them. An obliging communication from the clergyman of the parish, with a rough sketch, shows that the old dial is now in a most dilapidated condition. A large portion of the side on which the inscription was made has been broken off, and the only letters that remain are TE ME ME. On another side is the well known local name, "Christian," with the date below; and on the third face of the square stone on which the dial was placed, are the "three legs conjoined in the fesspoint," the arms of the Island.

The fourth side seems to be altogether broken away. "De : Whitehouse" may still be deciphered on the solid shaft.

The name of "Christian" with the date, which is thirty-one years after William Christian, *alias* William Dhône, or Fair-haired William, was executed under the *régime* of the famous Countess of Derby, has an historical interest connected with the Isle of Man; and this has been greatly enhanced by the Introductory Preface which Sir Walter Scott added to "Peveril of the Peak." Whatever disloyalty the family may have shown, the restoration of their property, and the fine imposed on the Derby estates, are facts which acknowledge that William, the Receiver General, had been unjustly put to death; and the member of the family who raised this dial, with its serious inscription, was probably both a wiser and sadder man than some of his ancestors had been.

We happen to have known intimately, in their old age, three sisters of the same maiden name and family, long years ago, whose parents had settled at Dunkirk in France. It was their relative, Fletcher Christian, who gained notoriety as leader of the mutiny on board the *Bounty;* when Lieutenant Bligh, who commanded, was turned adrift with a few companions in a boat on the great Pacific ocean. The hand which pens this has played in childhood with the bullet which Admiral Bligh wore in his old age on a blue ribbon round his neck, and which had been the weight he used in apportioning food to his fellow voyagers over 4,000 miles of sea in the open boat.

294.

TEMPUS AD LUCEM DUCIT VERITATEM.

Time brings truth to light.

Noted by Mr. Howard Hopley, but no place named.

295.

TEMPUS EDAX RERUM.

Time the devourer of (all) things.

This phrase from "Ovid's Metamorphoses" is on the plate of a pillar-dial in Easby Churchyard, which is picturesquely situated close to the ruined abbey by the river side, about a mile from Richmond, in Yorkshire. It is also on Dewsbury Church, with the date 1816; as well as on the porch of Gulval Church, in a pretty part of Cornwall, near Penzance, dated 1810. Time is here represented walking above the dial face, and holding both a scythe and hour glass. It occurs too on a house at Rye, where there is the additional inscription from Young's "Night Thoughts:"

> THAT SOLAR SHADOW, AS IT MEASURES LIFE,
> IT LIFE RESEMBLES TOO.—*Night* ii.

Quarles says, *Book* iii. *Emblem* 13,

> " Read on this dial, how the shades devour
> My short-liv'd winter's day; hour eats up hour;
> Alas! the total's but from eight to four."

296.
TEMPUS FUGIT.
Time flies.

Is on a florid dial-pillar in Handsworth Rectory garden, Yorkshire; also on the old church at Bridlington, on the Yorkshire coast. It is likewise inscribed on a marble dial at Kirk Braddan in the Isle of Man, with the date 1860. Over the dial face are the arms of the Island, three legs conjoined in the fesspoint, and beneath it are the Manx words—

> TA NY LAGHIN AIN MYR SCAA.

Our days are like a shadow.

At Ossington, Nottinghamshire, the seat of the Right Hon. J. E. Denison, the Speaker of the House of Commons, there still stands a dial bearing this motto. In "Thoroton's History of Notts," it may be seen figured in the same plate with the church and hall as they stood in the time of Charles I.; but both of these have since been rebuilt. The sun-dial alone remains. Two lines, from Virgil's third Georgic, 284-5, are on a house-dial at West Felton, Salop; the second of which is devoid of application.

> SED FUGIT INTEREA, FUGIT IRREPARABILE TEMPUS,
> SINGULA DUM CAPTI CIRCUMVECTAMUR AMORE.
>   —JOHN DEVASTON FECIT. 1789.

But meanwhile it flies, irreparable time flies;
Whilst we captivated by love pursue our several objects.

297.

TEMPUS LABILE.

Slippery time.

This is on a dial, facing south, over the kitchen-garden door at Esholt Hall near Leeds. There was originally a nunnery at "Esteholt," as Dugdale writes it, which was a cell to Sinningthwait, and of the Cistercian order. Pope Alexander III. took this nunnery into his protection in 1172. The present hall was built in the early part of the last century by Sir Walter Calverley, Bart., and in 1754-5 it was sold to Robert Stansfield, Esq., to whose representative, W. R. C. Stansfield, Esq., it lately belonged. The same motto is on an old, nicely-carved stone dial, which is fixed against the front of a cottage house in Bishopthorpe, near York; and below is a small, apparently marble slab, let into the wall, with the date 1691. They have possibly no connection with each other, and may be relics of some former archiepiscopal buildings.

298.

TEMPUS UT UMBRA PRÆTERIT.

Time passes by as a shadow.

One of the dial mottoes at Brougham Hall. See Nos. 202 and 332.

299.

TEMPUS VITÆ MONITOR.

Time the admonisher of life.

The locality of this has not been noted.

300.

TEMPUS VOLAT.

Time flies.

Is one of the mottoes suggested by the Rev. W. L. Bowles for "Morning Sun," and he thus renders the meaning:—

> "Oh, early passenger, look up—be wise,
> And think how, night and day, time onward flies."

See Nos. 65, 237.

301.

THE LAST HOUR TO MANY, POSSIBLY TO YOU.

On the church at Hartlepool, co. Durham.

302.

THE NATURAL CLOCKWORK BY THE MIGHTY ONE
WOUND UP AT FIRST, AND EVER SINCE HAS GONE:
NO PIN DROPS OUT, ITS WHEELS AND SPRINGS ARE GOOD,
IT SPEAKS ITS MAKER'S PRAISE THO' ONCE IT STOOD;
BUT THAT WAS BY THE ORDER OF THE WORKMAN'S POWER;
AND WHEN IT STANDS AGAIN IT GOES NO MORE.

JOHN ROBINSON, RECTOR.
A. DOUGLASS, CLERK FECIT. } A.D. 1773.

THOMAS SMITH
SAMUEL STEVENSON } CHURCHWARDENS.

SEAHAM IN LATITUDE $54^D\ 51^M$

On the south porch of Seaham Church, co. Durham. The motto is above the dial cut in a heavy stone of ungainly shape.

303.

THE NIGHT COMETH. 1800.

Over Melsonby church door in Yorkshire.

304.

ΘΕΟΣ ΓΕΩΜΕΤΡΕΙ.

God measures the earth.

MIhI DeVs LVX et saLVs.

God is my light and salvation.

On Hadleigh Church, Suffolk. In the second motto the larger letters form a chronogram, making the date 1627. This number is thus formed: MDLLXVVVII.

305.

THESE SHADES DO FLEE
FROM DAY TO DAY:
AND SO THIS LIFE
PASSETH AVVAIE.

In front of Marrington Hall, Shropshire, on the lawn, is a curious old four-sided dial, thus inscribed round the pillar, near the top. It is coeval with the house, and dated 1595. The shaft of the dial is set in a solid square stone at the base, round the chamfer of which runs the legend:—

FOR CHARITI BID ME ADW (ADIEU?) WHO WROUGHT THIS STONE FOR THEE TOMB OF R L L.

These letters are the initials of Richard LLoyd. On the sides of the shaft are various heraldic bearings, emblems, and devices—the arrow, death's head, cross bones, &c., mingled in arrangement, and showing the arms of families who have owned the property. Amongst these are those of Newton, to which family, it is believed, that Sir Isaac Newton belonged. In out-of-the-way corners of the stone there are many dials curiously inserted. The other inscriptions are:—

DEUS MIHI LUX.

God is my light.

FUI UT ES, ERIS UT SUM.

I was as thou art, thou wilt be as I am.

UT HORA SIC VITA.

Life is as an hour.

The dial seems to be the sepulchral monument of Richard LLoyd, either erected during his lifetime or placed over his remains.

On the lawn of the same mansion is another dial, bearing the date 1612, with the same initials R L L prettily interwreathed in the following motto:—

FINIS ITINERIS SEPULCHRUM.

The grave is the end of the journey.

306.

THUS ETERNITY APPROACHETH. G. HOLDEN, 1766.

There is a sun-dial over the south door of Pilling Church, Lancashire, which is thus inscribed. The Rev. G. Holden was the incumbent at the time when the dial was erected. Also on Thornton Church, in the same county.

307.

THUS THE GLORY OF THE WORLD PASSES AWAY. 1807.

Over the door of the church at Willerby, near Scarborough. The dial is plain and circular in form.

308.

THY DAYS ARE LIKE A SHADOW THAT DECLINETH.

Over the porch entrance of St. Madeon's Church, near Penzance. The situation is very fine, and commands a view of St. Mount's Bay. The churchyard is full of trees. "My days are like a shadow that declineth."—Psalm cii. 11. See No. 171.

309.

TIME AND TIDE TARRY FOR NO MAN.

This motto is on a dial in Brick Court, Middle Temple, which was repainted in 1861, and bears the initials J. T. A. In common with the other Temple dials the lamb and flag are represented on it. The motto is probably very old; and may be read with reference to the time when the lawyers went from their chambers to the courts at Westminster by boat; and the favouring tide in the river was an important element in conveying them in time for business. The same inscription is to be seen on the Hall in New Inn, Wych Street, London.

310.

TIME CAN DO MUCH. 1777.

Is on a dial in the garden of Leventhorpe Hall, near Leesd, the residence of J. Towlerton Leather, Esq.

311.

TIME FLIES. 1781.

On a white house near the wall of the sea at Hartlepool. In the sixth book of Wordsworth's "Excursion" there is a pleasant episode showing how two political opponents, "flaming Jacobite and sullen Hanoverian," used to meet and discuss in "The churchyard among the mountains;" and finally agreed to lie after death in one spot to be marked by a dial, which was thus inscribed:—

> TIME FLIES; IT IS HIS MELANCHOLY TASK
> TO BRING, AND BEAR AWAY DELUSIVE HOPES,
> AND REPRODUCE THE TROUBLE HE DESTROYS.
> BUT, WHILE HIS BLINDNESS THUS IS OCCUPIED,
> DISCERNING MORTAL! DO THOU SERVE THE WILL
> OF TIME'S ETERNAL MASTER, AND THAT PEACE
> WHICH THE WORLD WANTS, SHALL BE FOR THEE CONFIRMED.

It must be acknowledged that these lines partake a little of the prosiness of which the poet is so often accused.

312.

TIME IS, THOU HAST: SEE THAT THOU WELL EMPLOY;
TIME PAST IS GONE: THOU CANST NOT THAT ENJOY.
TIME FUTURE IS NOT, AND MAY NEVER BE;
TIME PRESENT IS THE ONLY TIME FOR THEE.

Over the door of a schoolmaster's house at Leyburn, Yorkshire. Another version gives the two first lines thus:—

> "Time was, is past: thou canst not it recall;
> Time is, thou hast: employ the portion small."

313.

TIME IS ON THE WING, AND THE MOMENTS
OF LIFE ARE TOO PRECIOUS TO BE
SQUANDERED AWAY ON TRIFLES.

At Hesketh, in Lancashire, is a dial with two mottoes. The first is that just given, and the other has been already quoted, see No. 9:—

> AH, WHAT IS HUMAN LIFE, &c.

314.

TIME WASTED IS EXISTENCE, USED IS LIFE. 1826.

Over the porch entrance of the church at Hutton Buscell, Yorkshire, on an oval-shaped dial, is this inscription. The same idea is expressed by Herrick.

> "Long have I *lasted* in this world 'tis true,
> But yet these years that I have *lived*, but few.
> Who by his grey hairs doth his lustres tell,
> Lives not those years, but that he lives them well.
> He lives, who lives to virtue, men who cast
> Their lives to pleasure do not *live*, but *last*."

315.

ΤΟ ΣΗΜΕΡΟΝ ΜΕΛΕΙ ΜΟΙ,

ΤΟ Δ'ΑΥΡΙΟΝ ΤΙΣ ΟΙΔΕ;

To-day concerns me, but of to-morrow who knows?

Inscribed on the base of the pedestal of a pillar dial at Watton Abbey, near Driffield.

316.

TRIFLE NOT, YOUR TIME'S SHORT. 1775.

At Milton, near Gravesend. So says Sir Walter Scott:—

> "Nay, dally not with Time, the wise man's treasure,
> Though fools are lavish on't—the fatal Fisher
> Hooks souls, while we waste moments."

317.

**TRUE AS THE DIAL TO THE SUN,
ALTHOUGH IT BE NOT SHONE UPON.** 1808.

The lines are from "Hudibras," and the dial is placed under the eaves and close to a buttress of Halifax parish Church, Yorkshire. It is horseshoe shaped, and has gilt lettering on a black ground.

318.

TUA HORA RUIT MEA.

Thine affairs my hour overthrows.

This is on a dial in the cloister of the quiet old Franciscan monastery, which is picturesquely perched on the top of one of the olive-clad hills behind Nice. Within the cloister is a garden of early flowers, which the monks cultivate and present at the wicket gate to lady visitors who may not enter their retreat. The Latin of the motto is monkish, and the translation doubtful. *Ruo* is treated as an active verb, and the dial, as usual, is supposed to speak.

319.

ULTIMA FORSAN.

Perhaps your last hour.

This motto was observed in Switzerland. The dial was round in shape and modern; it faced south and was painted blue and grey on a wall.

320.

ULTIMA LATET.

Your last hour is hidden.

At an angle of the cloister beneath the belfry of the Franciscan monastery, near Nice, which has been already mentioned, No. 318. Mr. Hopley read this inscription, and says that "when the old monks tolled the Angelus, this dial was half in gloom, and the evening hours were shrouded in shade."

321.

ULTIMA NECAT.

The last hour kills.

At Spotorno.

322.

ULTIMAM TIME.

Fear your last hour.

We have no note of the place.

323.

UMBRA DEI.

The Shadow of God.

On the cross-dial at Elleslie, near Chichester.

324.

UMBRA LABITUR, ET NOS UMBRÆ.

The shadow glides away, and we are shadows.

One of the mottoes on Glasgow Cathedral. See Nos. 75, 216.

325.

UMBRÆ TRANSITUS EST TEMPUS NOSTRUM.
S. Sykes, Fecit. Decem. 22, 1790.

Our time is the passing away of a shadow.

On a house dial in Wentworth, Yorkshire; also at Cuers in France.

326.

UMBRAM DUM SPECTAS REFUGIT REVOLUBILE TEMPUS.

Whilst you view the shadow the revolving time flies back.

In Alderley churchyard, Cheshire.

327.

**UNA DI QUESTE T' APRIRÀ LE PORTE
DI VITA LIETA, O DI SPIETATA MORTE.**

*One of these hours shall open thee the gate
Of blissful life, or of relentless fate.*

Between Nervia and Convento, on the Riviera. Dean Alford read and translated this motto.

328.

UNA UMBRA ET VAPOR EST HOMINUM VITA.

Man's life is a shadow and smoke.

There is a curious device on this sun-dial, which stands in Helston churchyard, Cornwall. It represents St. Michael, we presume, as the figure is robed, winged, and with rays of glory round the head, standing betwixt two towers, and driving his spear into a dragon which is at his feet.

329.

UNA HARUM VITÆ HORARUM ERIT ULTIMA. 1814.

One alone of these hours of thy life will be the last.

On a church near Queen Hortense's Chateau of Arenemburg. The dial is circular in form. Time with his scythe is in the centre, and over him, a sun, from which issues the gnomon.

330.

UNAM TIME.

Fear one hour.

On a house at St. Pierre, on the Great St. Bernard road. For a description of the situation see No. 107.

331.

UNAM TIMEO.

One hour I fear.

On the church of Stassano in Piedmont.

332.

USQUE HUC CRESCIT.

Even so far there is increase.

This is at Beziers in France.

333.

UT HORA SIC VITA.

Life is as an hour.

This motto may be seen on a little white wooden dial over the porch of the church at Stanhope, in the county of Durham. In Adel Churchyard, near Leeds, the same inscription may be read on a pillar-dial which is octagonal in form. The words are written on a scroll on the dial-plate, which also contains the maker's name: "J. Mann Ebor. fecit ex donatu 1682." It is also one of the mottoes at Brougham Hall, see No. 202 and 298, as well as on Felton Church, Northumberland, with the date 1724, and on Ormsby Church, Yorkshire, 1776.

334.

UT UMBRA DECLINAVERUNT.
They have gone down as a shadow.

This may be read at Trafiume, near Cannobio, Lago Maggiore.

335.

UT UMBRA SIC VITA.
Life is as a shadow.

This motto, dated 1695, is on a dial at Morden College, Blackheath; also at Morvah Church, West Cornwall, with the date 1-29 partially defaced. It is engraved, too, on one of the four corner pinnacles of the churchyard wall at Sleights, near Whitby. Here the motto is below the dial, which faces south. On the east side of the same pinnacle is another dial, with the date 1761, and the initials R T B and G. B. It was in this year that Robert and Tabitha Bower built the church. It may be read at Shaftesbury, Dorset; and on a dial over the south porch of Hartest Church, Suffolk. On the terrace at Derwent Hall, in Derbyshire, now the property of the Duke of Norfolk, the same inscription occurs with the additional line :—

ADVENIET ILLA DIES : SEMPER PARATUM.

That day will come: (the dial) is always prepared.

The word "horologium" must be understood.

336.

UT UMBRA, SIC VITA TRANSIT.
Life passes away like a shadow.

This is on a glass dial in a window of Election Chamber, Winchester College. The shape of the dial is an oblong square, set in an oval frame of richly-coloured glass. The figures are in a yellow border. The motto is on a scroll in the centre of the upper half of the pane which forms the plate, and at one corner is the mysterious fly already noticed, see No. 62. Henry King, writing in the 17th century, says:—

> "What is the existence of man's life?
> *   *   *   *   *
> It is a dial which points out
> The sunset as it moves about:
> And shadows out, in lines of night,
> The subtle changes of Time's flight;
> Till all obscuring Earth hath laid
> The body in perpetual shade."

337.

UT VITA SIC FUGIT HORA.

The hour passes away like life.

On a Sanctuary dedicated to St. Francis of Assisi, which is situated on a hill that rises above the lake and town of Orta.

338.

UT VITA SIC UMBRA. 1833.

Life is as a shadow.

On a house at Kirby Moorside, Yorkshire. The dial is plain and circular in shape.

339.

VÆ TERRÆ ET MARI, QUIA DESCENDIT DIABOLUS AD VOS, HABENS IRAM MAGNAM, SCIENS QUOD MODICUM TEMPUS HABET. APOC. C. XII. V. 12.

Woe to the earth and to the sea, because the devil is come down unto you, having great wrath, knowing that he hath but a short time.

The locality in which we find this inscription, which is from the Vulgate translation, Douay version, is very interesting. The dial, which seems to be held by an eagle in a very startled condition, is over an archway which leads into the great convent square at the top of the Sacro Monte at Varallo in Piedmont; and through this opening pilgrims from all parts of Italy have been accustomed to pass and repass in order to pay their devotions at the " Nuova Gerusalemma nel Sacro Monte di Varallo."

The Sacro Monte was founded in 1486 by Bernardino Caimo, a Milanese nobleman, and it grew rapidly in riches and reputation; the visits paid to it by Archbishop Carlo Borromeo contributing no little to its renown. Forty-six chapels or oratories are dotted over the hill, in each of which there is a scene from the life of our Lord, represented by groups of life-sized terra-cotta figures, that are clothed and painted in imitation of the reality; whilst the walls are covered with frescoes, on which the Alpine artists have exercised their skill for many years. Among these are some of Gaudenzio Ferrari's finest works, but the screens and partitions which enforce that distance which "lends enchantment to the view" of these figure groups, are by no means favourable to an examination of the frescoes. There are, however, some very striking groups, notwithstanding the drawbacks of age, eccentricity, and excessive realism. Sometimes a grand force and truth of expression are revealed, which must have made the sacred scenes come home to the hearts of the mountaineers. The Sacro Monte is crowned by the Convent, which overlooks the lovely Val Sesia, whilst the town of Varallo lies at the foot of its Mount Calvary.

The dial is large, painted on the wall, and much ornamented. A kind of eagle's head and wings rises above the plane, and something of the same sort appears below; the whole being enclosed in a narrow border. The width of the dial exceeds that of the arch beneath. The lines on the face show the Italian hours only,—from i to xxiv. The tropics of Cancer and Capricorn are described upon it, and the parallels of the sun's course at his entrance into the twelve signs of the Zodiac, together with the characters of the signs. The motto is on a spiral scroll on one side of the dial, and a corresponding scroll on the other side, somewhat defaced, has an imperfect Latin inscription, which refers to the mechanism of the dial. This is preceded by the figures -645, which may be a partially obliterated date of 1645, for it is evident that the dial must be tolerably old from the fact that only the Italian hours are recorded.

340.

VASSENE 'L TEMPO, E L' UOM NON SE N' AVVEDE.—
DANTE, *Purg.* c. IV.

Time flies, though man perceives not it hath been. (Wright's translation.)

This is on a house in the Via Brondolo, Padua; and "Padova la dotta" may be said to maintain its character for learning, even in its dial; and to show its fidelity to the memory of Dante, who is reported to have lived there in 1306. Nevertheless, the present dial is modern, and is painted on a wall in a back street—south, declining east—just above the green shutters of the windows of the first story. The round arched doorway below, supported by a pillar on each side, opens into a carpenter's shop; but there lingers an air of departed grandeur about the building, which suggests that its owners in the last century were people of greater consequence than the present occupiers.

341.

VEILLEZ SUR TOUTE, CRAIGNEZ LA DERNIERE.

Watch over all, fear the last.

This inscription is on a scroll over a dial which is painted on the wall of a house at Cannes, belonging to M. Negrin, Notaire, as a board below informs us.

342.

VENIO UT FUR.

I come as a thief.

Copied by Mr. Howard Hopley, but no place named. In Shakespeare's 77th sonnet we read:—

"Thou by thy dial's shady stealth may'st know
Time's thievish progress to Eternity."

343.

VERA LOQUOR AUT SILEO.

I speak the truth or am silent.

On Highgate School.

344.

VESTIGIA NULLA RETRORSUM.

There are no steps backward.

This is on one of the plain business-like dials of the Temple, and was restored in 1861 with the initials J T A upon it. The same motto is on a house at Brompton-on-Swale, a village near Richmond in Yorkshire, where Miss Carter formerly lived, the sister of one of the prebends of Lincoln, who probably suggested the words.

345.

VIA CRUCIS VIA LUCIS.

The way of the Cross is the way of light.

At Hurstpierpoint School there is a dial, shaped as a recumbent Cross, which bears this inscription. The hours are indicated by the position of the shadow on the different points of the Cross. See No. 122.

346.

VIA VITÆ.

The way of life.

Over a large square stone dial which was placed between pinnacles against the south side of the tower of Sheffield old Parish Church. The dial was removed when a new clock was erected, but happily the Vicar has taken care to have the older time-keeper restored to very nearly its former position. The same inscription is on the Church porch of Himbleton in Worcestershire, and on the cross-dial at Elleslie, near Chichester.

347.

VIDE, AUDI, TACE.

See, hear, and say nothing.

The position of this dial motto is not identified.

348.

VIDES HORAM ET NESCIS FUTURUM. 1836.

You see the hour, and know not the future.

The dial is a large erect one, painted south, declining west, on the wall of a house at Pra on the Riviera, which forms part of what was once a little chapel, the tower and bell of which still remain. The small adjoining house was probably the priest's residence. The belfry tower is the oldest and most picturesque portion of the building, constructed in the true Genoese style, with alternate stripes of black and white marble. All other traces of its former use have now disappeared. The windows on either side of the dial may usually be seen festooned outside with clothes after a washing day, and the tenants are poor people. It stands in the middle of the Piazza, which is the great rendezvous of all the inhabitants of Pra. This place is a large village, seven or eight miles west of Genoa, where both fishing and boat-building are carried on.

349.

VIGILATE ET ORATE.

Watch and pray.

Over the south porch of Rothwell Church, near Leeds, with the further inscription, "I. Verity fecit. Lat. 53.15." The motto and figures are on a blue border.

350.

VIGILATE ET ORATE: TEMPUS FUGIT. 1781.

Watch and pray : Time flies.

High up on the tower of Ellastone Church, near Ashburne, Derbyshire, is this motto. The dial is fixed over a small built-up window, or what seems like it, and below is the additional inscription, KNOWE THYSELFFE, which looks of older date.

351.

VIGILATE, QUIA NESCITIS HORAM.

Watch, for ye know not the hour.

The dial, defaced and without date, is painted between the windows of a house at Arles, having curious balconies, and which looks into a small square where the peasants on *fête* days dance their national Catalan dance in the white caps and *espartillos* that are still worn in the Eastern Pyrenees. For this Arles must not be confounded with Arles in Provence; but is a town with about 2,000 inhabitants near the head of the valley of the Tech, thirty-nine kilometres from Perpignan, and it is surrounded by mountains of which the Canigou towers highest. It has an ancient church, with a cloister of great elegance and beauty, dating from the thirteenth century, and is about two or three hours distant from the Spanish frontier.

352.

VIRTUS AD ASTRA TENDIT, IN MORTEM TIMOR.

Virtue tends to the stars, fear to death.

This is on the Chateau of Oberhofen, which is situated on the Lake of Thun, and is a picturesque building, close to the water's edge, belonging to the Count Pourtalès.

353.

VITA FUGIT VELUT UMBRA. 1790.

Life flies as a shadow.

On the church at Sandal, Yorkshire.

354.

VITA HOMINIS SICUT UMBRA FLUIT.

The life of man flows away like a shadow.

At Courmayeur.

355.

VITA QUASI UMBRA.

Life is as a shadow.

At Sproughton Rectory, near Ipswich.

356.

VITA SIC TRANSIT. 1817. Lat. 54·20 N. S. 20 W.

W. Putsey delineavit.

So life passes away.

This is on a square dial on Pickering Church, Yorkshire.

357.

VITA UMBRA.

Life is a shadow.

On Archbishop Abbot's Hospital, Guildford. The dial is nearly at the top of the entrance tower.

358.

VIVE MEMOR QUAM SIS ÆVI BREVIS. 1767.
JOHN METCALFE, FECIT., CHURCHWARDEN. 1767.

Live mindful how short-lived thou art.

On a dial fixed outside the terrace walk at Wentworth Woodhouse, Yorkshire, the seat of Earl Fitzwilliam, K.G. The same motto is on the noble owner's glass house, with "Lat. 53-28. Dec. W. 34° delineavit Johan. Metcalfe, 1766;" and on the Lodge there is a dial inscribed, A. D. 1764. OMNIA SUNT HOMINUM TENUI PENDENTIA FILO; *all human affairs hang on a slight thread.* On a house dial in Wentworth is the motto already given, No. 5. AB HOC MOMENTO, &c., DEC. 26. W. LAT. 53·27. N. DELINEAVIT JOHAN. METCALFE 1765.

359.

VIVIT MEMORIA ET FUGIT HORA.

Memory lives and the hour flies.

On a house at Monthey, in Canton du Valais, Switzerland. The dial is very plain, and circular in form.

360.

VIVITE, AIT, FUGIO.

Live, it says : I fly.

This inscription is thus alluded to in a letter from Bishop Atterbury to Pope, dated Bromley, May 25, 1712: "You know the motto of my sun-dial, *Vivite, ait, fugio.* I will, as far as I am able, follow its advice, and cut off all unnecessary avocations and amusements." In the same correspondence of the Bishop the following epigram occurs:—

> " Vivite, ait, fugio.
> Labentem tacito quisquis pede conspicis umbram,
> Si sapis, hæc audis : ' Vivite, nam fugio.'
> Utilis est oculis, nec inutilis auribus umbra;
> Dum tacet, exclamat, ' Vivite, nam fugio.' "

The dial was probably a mural one on the ancient moated palace of the Bishops of Rochester, at Bromley, which was pulled down by Bishop Thomas in 1774. The building which he substituted has ceased to be an episcopal residence.

361.

VOICI VOTRE HEURE.

Behold your hour.

To be read near Geneva.

362.

VOLAT SINE MORA.

It flies without delay.

In the cloister of the Franciscan monastery behind Nice. See Nos. 318 and 320.

363.

VULNERANT OMNES, ULTIMA NECAT.

All wound, the last kills.

Painted, south declining east, on the church tower of Urugne, department Basses Pyrenées. The motto is in large capitals above the dial, which is placed below the open arch of the belfry.

Urugne is on the great western road leading from France into Spain, and is the last French post station, having the wild irregular ridges of the Spanish mountains in full view. The dark Spanish-looking church has associations connected with the Peninsular War. The "Subaltern" gives an account of a night spent in it after the assault and capture of the village on the previous day, in November, 1813, when he and his men were cantoned in the church, whose thick walls were proof against the field artillery of the French. This village formed part of Marshal Soult's famous position in front of St. Jean de Luz.

The same motto is also at Cawder, near Glasgow.

364.

WACHET; DENN IHR WISSET NICHT, UM WELCHE STUNDE EUER HERR KOMMEN WIRD.

Watch, for ye know not at what hour your Lord doth come.—MATT. xxiv. 42.

At Erstfelden, near Altdorf, in the Canton Uri, there is a dial painted on the wall of a little village church. It is circular in form, with the face of the sun at the top, out of which comes the gnomon. A full-length skeleton stands on each side of the dial, like the supporter to an heraldic shield, and appears to hold up the plate. Beneath it are cross-bones, and the motto is above. Some words under the cross-bones are defaced and illegible.

Erstfelden is a small village in the valley of the Reuss, where the road over the Surenen Pass joins that of the Great St. Gothard.

365.

WATCH AND PRAY. 1735.

On Alwalton Church, Huntingdonshire.

366.

WATCH AND PRAY,
TIME HASTES AWAY.

Over a cottage at Barton, near Darlington.

367.

WATCH AND PRAY,
TIME PASSETH AWAY LIKE A SHADOW.

On the church at Isleworth, Surrey, which was built in 1705. The dial is quaint. Above it Time is represented as a bearded old man with wings, and he reclines on his back holding a scythe, the point of which touches a scroll at his feet, on which is written "Watch and pray;" whilst "Time passeth," &c., is on another scroll above the dial face, which half hides a radiated sun, and from this comes the gnomon. The dial is lineated, and the hours marked at several distant places, such as Jerusalem, Moscow, &c.

368.

WATCH, FOR YE KNOW NOT THE HOUR. Inscribed 1862, M. G.

The fine old village church at Ecclesfield, in Yorkshire, was externally rebuilt *circa* 1470. In the middle of the churchyard, on the south side, stands the broken shaft of a mortuary cross. Two stone steps form the pedestal which supports it, and it is surmounted by a small copper dial. On the upper step, around the shaft, Mrs. Alfred Gatty had the above motto engraved in 1862.

369.

WATCH WEEL.

This, which is the heraldic motto of Sir Walter Scott, is inscribed on a very graceful pillar dial, having four faces to the different points of the compass, which stands in a little shrubbery near the arch of the ruined Abbey of Dryburgh, under which lie the mortal remains of the great romance writer and of Lady Scott. It is supposed that the date of the dial is 1640, and that the Haliburtons found it and reared it in its present place.

Dryburgh Abbey belonged to the Haliburton family before it came into the possession of the Earls of Buchan; and Robert Haliburton, grand-uncle of Sir Walter Scott, had settled it by will upon the poet's father, as heir in the maternal line. But this ancient patrimony was lost to the Scotts through Mr. Haliburton's commercial adventures; "and thus," wrote Sir Walter, in his brief autobiography, "we have nothing left of Dryburgh, although my father's maternal inheritance, but the right of stretching our bones where mine may perhaps be laid ere any eye but my own glances over these pages."

SUN-DIALS.

The collector sketched the dial on the 10th August, 1839, and thus wrote:—

> "'Watch weel,' lest thieves should enter while ye sleep—
> But pray to God His favour to obtain:
> Except the Lord Himself the city keep,
> The careful watchman waketh but in vain."

370.

WE RESEMBLE THE SHADOW. 1812.

On Wragby Church, Lincolnshire.

371.

WE SHALL —— 1693. (*scil.* DIAL *i. e.* die-all).

This somewhat cumbrous joke is not uncommon. It may be read in Buxted churchyard, near Uckfield, Sussex, where it is inscribed on an old and rather elaborately engraved dial. Also over the south porch of Bromsgrove church, in Worcestershire, is a dial inscribed *We shall* in old English characters. At Kedleston, in Derbyshire, it is *We must;* and on a house at Easton, near Stamford, there is *Wee shall* —. An old story connected with this quaint conceit is, that a certain pious cleric, who had seen the inscription "We must" on a sun-dial, and ascertained how the "die all" to conclude the sentence was obtained, ordered the words "we must" to be inscribed on the clock face of his church!

372.

WE SHALL DIE ALL.

On Walgrave Church, Northamptonshire.

373.

WHO DULY WEIGHS THE HOURS. 1715.

Is at Breage, in Cornwall, where Mrs. Godolphin was buried. This eminent lady, whose life was written by John Evelyn, one of her most intimate friends, was a daughter of Colonel Blagge, and born in 1652. She became a maid of honour to Queen Catherine at the court of Charles II., and in 1675 married Sidney, third son of Sir Francis Godolphin. What especially distinguished her was her pious, modest, and discreet character, whilst living at a court where Christian virtues were strange. She died in 1678, and was buried at Breage, where her husband's family had been settled before the Conquest. Her surviving husband was created Earl of Godolphin, and through their grand-daughter they are ancestors of Godolphin Osborne and the Duke of Leeds.

374.

WORK TO-DAY, AND PLAY TO-MORROW.

On Turner's Hospital, at Kirkleatham, near Redcar, Yorkshire. See also No. 73. Both the dials are plain in pattern, and circular.

375.

WORK WHILE IT IS DAY. Lat. 53′ 28′. J. S. 1849.

This sentiment is taken from St. John ix. 4, and is the motto on a dial placed on the south porch of Bradbury Church, Cheshire.

376.

YET A LITTLE WHILE IS THE LIGHT WITH YOU: WALK WHILE YE HAVE THE LIGHT. John xii. 35. 1671.

At Aynho. See 72.

377.

YOU MAY WASTE, BUT YOU CANNOT STOP ME.

Is painted on a board over the door of a chapel of ease on the Parade at Tunbridge Wells. There is no date, but the maker's name is recorded, "Alexr. Rae fecit." It has been written:

> "Time wastes us, our bodies and our wits:
> And we waste Time, so Time and we are quits."

Richard II. thus soliloquizes in his dungeon in Pomfret Castle:

> "I wasted time, and now doth time waste me.
> For now hath time made me his numb'ring clock:
> My thoughts are minutes; and, with sighs, they jar
> Their watches on to mine eyes, the outward watch,
> Whereto my finger, like a dial's point,
> Is pointing still, in cleansing them from tears.
> SHAKESPEARE, *Richard II.*, Act v. Sc. 5.

FURTHER NOTES ON REMARKABLE SUN-DIALS.

ALTHOUGH it was the interest attached to sun-dial mottoes which gave rise to this collection, it was seen after a time that the dials were interesting on their own account, from the variety and ingenuity of their structure; and though no inelegance of form or poverty of material can render a beautiful motto insignificant, it is a fact that some of the finest and most costly dials are without mottoes. Of such of these as have not been noticed before, we think that a brief description will be acceptable. The Greek-Egyptian dial, which was found at the base of Cleopatra's Needle, at Alexandria, in 1852, by J. Scott Tucker, Esq., and is now in the British Museum, is not only the most ancient form of dial, but it embodies the principle on which all other dials have since been constructed. The shape is that of a hemicircle, scooped into a square block of stone—in fact, a wooden bowl, if sawn across, would accurately represent, in its hollow, the very form which Berosus, the Chaldæan astronomer, is supposed to have introduced into Greece; whilst the steps which were found below this Greek-Egyptian dial seem to identify it with the Assyrian form adopted by Ahaz.

This mode of obtaining the shadow within a hollow has been common in more modern dials; notably so in a fine specimen which stands picturesquely near the side entrance to Madeley Hall, county Salop. No history attaches to the dial; but the mansion was formerly the country house of the prior of Wenlock Abbey. This remarkable dial is in form a solid cube of stone, standing on four low feet, and elevated on a platform of three steps. In each of the four sides is a hollowed-out circle, to receive the rays of the sun and by a gnomon record the hour; whilst around each of them are smaller hollows, round, triangular, and diamond shaped, which equally serve to indicate the hour at one time of day or another. The whole is very substantial and handsome, and it has a rounded top.

Perhaps the most elaborate, complicated, and beautiful dial in the world, is that of Buen Retiro, Churriana, near Malaga, in Spain, which is made of white marble, includes three stages in its construction, and is so carved and diversified in shape, especially in the upper part, that it contains 150 dials in all. The upper face includes a star, and there is another below, also a cross, which is inclined and covered with dials; and beneath them are engraved the two castles and lion, "Castilla y Leon," the royal badge of Spain. The sides of this front are hollowed and lineated, after the manner of the Greek or Assyrian model. On one side of the middle stage is a shell carved, which contains a dial, and another is beside it; and so ingeniously are the angles cut throughout this piece of machinery, that they do not seem to be indicated on the surface, but dials exist wherever there is room to scratch a line. The step on which the whole rests is formed of thin flat bricks, the pavement being black and white. It stands beside a stone tank, on a terrace which faces a lovely view over a fruitful plain and distance. The neighbouring hills glow in the sunlight, the sombre cypress trees cast their gloom around; and the melancholy glance of Time seems to be present, throwing its shade over its own fleeting footsteps, as these are expressed by the many gnomons on this remarkable instrument.

The white marble of the structure strongly contrasts with the dark sad green of the funereal trees; and as among the devices cut in the sides of this structure are the cockle-shell or scallop of the pilgrim, the star of hope, and the Cross of Christian faith, in contrast with the ducal crown, the cardinal's hat, and the kingly quarterings; enough, and more than enough, is suggested for serious meditation to anyone who visits this curious time-reckoner.

A small dial, cut on the same principle as that at Malaga, is in the Museum at Clermont-Ferrand, in Auvergne. It is made of white marble, but the lower half has been coloured a brick red, with a star painted upon it. The top is a hollow globe, set, as it were, in a cup of larger size, upon the rim of which the hours are marked by the shadow of a gnomon. On the pedestal are various hollows and curves, as well as flat and perpendicular surfaces, on each of which one or more dials are traced to the number of about thirty. The whole block is not more than a foot and a half high, and is said to date from the 16th century. It was found at the Chateau of Tournouelles in Auvergne.

There is an upright dial of white marble, projecting from and at right angles with the façade of the Church of Santa Maria Novella at Florence. It was placed there by the Grand Duke Cosmo I., as the inscription shows: "Cosm. Med. magn. Etr. dux, nobilium artium studiosus, astronomiæ studiosis dedit, anno D. MDLXXII. *Cosmo Medici, the Grand Duke of Etruria, a student of the ennobling arts, gave this to the students of astronomy,* A. D. 1572. The dial was the work of Fra Ignazio Danti of the Dominicans, to which order the Church and Monastery belonged. On the other side of the slab may be seen the "armilla di Tolomeo"—*sphere of Ptolemy*—for observing the ingress of the sun into the first point of Aries.

The Church of Santa Maria Novella is the one which Michael Angelo called "his Bride." Within the church there is a clock in one of the transepts which bears the following inscription :—

> Sic fluit occulte, sic multos decipit ætas;
> Sic venit ad finem quidquid in orbe manet.
> Heu! heu! præteritum non est revocabile tempus;
> Heu! proprius tacito mors venit ipsa pede.
>
> *So flows the age unperceived, so it deceives many;*
> *So comes to an end whatever remains in the world.*
> *Alas, alas, the time past is not to be recalled;*
> *Alas, death itself comes nearer with silent foot.*

Something more may be stated of "Queen Mary's Dial," at Holyrood than was given in the Introductory Chapter; also of Scotch dials generally.

The former was unnoticed for many years, both by residents and tourists, because it stood in the gardens of Holyrood Palace; and these were leased out to a market gardener, who turned them to profit. Queen Victoria's visit to this interesting spot brought it again into due prominence, and it now stands an object of historical interest. But the unhappy Mary Stuart never saw it; for accounts, recently discovered in the Register Office, show that

£408 15s. 6d., Scots, were paid by King Charles I. to "John Mylne, Maisoune," on account of "the working and the hewing of the dyell." The monograms and other devices carved upon it testify also to its belonging to this period, and it is said to have been presented by Charles I. to his Queen.

At Cramond House, near Edinburgh (C. C. Halket Inglis, Esq.), is a pillar dial, of far more delicate and complicated construction. All the gnomons, which are many in number, are on the top of the pillar, which is very beautifully ornamented. At Midmar Castle, Aberdeenshire, (John Gordon, of Cluny, Esq.,) is a more simple and solid structure, after the fashion somewhat of that at Holyrood. It is four-faced.

At Kelburne House, Ayrshire, (the Earl of Glasgow) is a tapering pillar, based on steps, and bulging in the centre, so as to offer three ledges on which dials are closely crowded. The whole pillar is completely covered with variously-shaped sinkings—heart, cross, shell, &c.—besides the common flat faces for the gnomon. At the top of the column is a handsome vane, surmounted by a thistle and leaves.

At Dundas Castle, Queensferry (James Dundas, Esq., of Dundas), at Inch House also, (W. J. Little Gilmour, Esq.), at Meggetland (Miss Sievewright), at Woodhouselee (J. Stuart Tytler, Esq.), at Carberry Tower (Lord Elphinstone), all in the neighbourhood of Edinburgh, there are fine old dials. At Heriot's Hospital, Edinburgh, there is a dial of the same character, and at Innes House, Morayshire (the Earl of Fife).

At Rubislaw Den, near Aberdeen, there is a sun-dial, which has been for several generations in the possession of the family of Skene, of Rubislaw, and has been considered a good specimen of the structures of the period. A sketch of it by the late James Skene, Esq., was kindly promised by the family, for inspection, but could not be found.

At Craigmillar (W. Little Gilmour, Esq.) is another dial of the same type: and at Leuchars, (Sir Coutts Lindsay, Bart.,) is a modern dial, designed by Lady John Scott, who also gave the pattern of that at Cawston, near Rugby. At the base of this fine dial, which is fashioned after that at Holyrood, are the arms of Scott and Spottiswoode impaled: with their respective mottoes, "Best riding by moonlight," *Patior ut potiar*; also "Bellenden!" the Scott slogan, ("Their gathering word was Bellenden"), *Amo*, the Montagu motto, and "John and Alice Scott" all in compartments. The pillar is encircled by a serpent. Above are the crests of the two families, with the initials of Lord and Lady John Scott; also "United in time," "Parted in time," "To be reunited when time shall be no more." These are also in separate compartments. The ball at the top is sliced over, triangular-wise, for the dial planes. Two steps support the structure. Indeed, at most gentlemen's seats in Scotland there are sun-dials. At Bowland, for instance, near Galashiels (Wm Stuart Walker, Esq.), there are two solid stone side posts to what had once been a gateway entrance, and on the tapering top of each is a globe, round which the hours are figured; whilst the gnomon is an iron rod pointing from the north pole. A sketch is before us, made in 1839—Eheu, labuntur anni! They are common, too, in the villages of Scotland, and one is cut on a buttress, near the south door of Melrose Abbey, dated 1661.

The dial at Pinkie House, Musselburgh, the residence of Sir Archibald Hope, Bart., is supposed to have been put up by Alexander Seton, Earl of Dunfermline, who died there in 1622. It rests on a wall, is four-sided: and a carved pinnacle rises from the square, that is much carved, and has a weathercock at the top.

There is a sun-dial in the garden at Northbar, the property of Lord Blantyre, near Erskine, which is his lordship's seat. It consists of a female figure carved in stone, and bearing a dial on her head. The date upon it is 1679, and the letters are D.M.G., being the initials of Donald MacGilchrist, to whom the place formerly belonged. It is an old-looking, quaint piece of sculpture.

At Polton, near Edinburgh (Mrs. Dundas Durham), are two dials which date about the end of the seventeenth century. One shows a figure of Time in relief, winged, holding a scythe, and supporting a globe on his knee: a square dial face is below. The other is a fragment. On a square base rests an hexagonally carved stone: on the face of both there have been dials; but how the structure terminated above is not known.

There are two curious old dials standing in Elmley Castle Churchyard, near Pershore, Worcestershire: one of which is placed on the eastern side of the burial ground. The stone of this is 1 foot 10 inches square, and is bevelled off at the shoulder, where it is surmounted by a globular-shaped top, that is covered with sinkings of various forms — hexagonal, pentagonal, heart-shaped, but chiefly circular. All of these are much weather-beaten and worn. In several of the indentations there remains a thin iron rod, which was the original gnomon; but many of them are beaten flat upon the stone. The whole height of the dial, not much exceeding three feet and a half, has caused it to be within reach of the children, who have loosened the globular top; and it now revolves, when pushed, on the iron spindle in the centre, which holds it to the lower stone. On the flat surfaces of the stone the former existence of twelve gnomons may be traced by the remains of the lead with which they were fixed. Across two of the hemispherical sinkings an iron rod extended diagonally; and two others contain thin metal gnomons, which are still tolerably perfect. When an examination and sketch of this dial were made, about nine inches of soil had to be cleared away from the base.

The other dial stands near the north-west angle of the churchyard, and is erected on a portion of the base and one of the steps of the old mortuary Cross. On this foundation there are six courses of stone masonry, rising to two feet six inches in height; and above them is reared a stone, so similar in size and shape to the dial already described, that one cannot help supposing they originally formed one structure. Three of these sides are marked with variously-shaped sinkings; and on the north side is a large shield, bearing the arms of Savage, with numerous quarterings. The manor of Elmley was granted by Henry VIII. to Christopher Savage, and this family has held property in the parish until within the last few years. There can be no doubt that this dial was put up at the cost of one of the family. At the top is a more modern four-faced dial, about ten inches square, sloped above like a house roof, which tells the hours in the ordinary way.

Of the few remaining Saxon dials, that which is over the south porch of the ancient church at Kirkdale, near Kirkby Moorside, seems to be the most remarkable. Old as the church is, and bearing evident traces of early ante-Normanic work, the dial is of still longer antiquity than the greater portion of the building, and has probably been moved and preserved in its present position in the repairs and alterations of the original edifice. The whole dial, with its inscriptions, measures seven feet in width by nearly two feet in height: in the centre is the dial itself, lineated, but not numbered, and shaped as a half circle. On the left side, as you face it, the following is engraved: "✠ Orm . gamal . suna . bohte . scs . Gregorius . Minster . donne . hit . wes . æl . to . bro." On the right side it proceeds: "can . & . to . falan . he hit . let .

macan . newan . from grunde . X. R. E & scs . Gregori . us . in . Eadward . dagum . cng . in . Tosti . dagum . eorl." This has been translated thus: "Orm, Gamal's son, bought S. Gregory's Minster, when it was all broken and fallen. He let it to be made new from the ground—to Christ and S. Gregory in Edward's days the king: in Tosti's days the Earl."

The story of the dial is further given by words engraved round and beneath the sun-dial itself. First we read: "✠ Dis . is . dæges . Solmorca . æt . ilcum . tide . ;" whilst below is: "✠ & Haward . me . wrohte . & . Brand . P. R. S. ;" which means, "This is the day's sun-marker at every season: and Hawarth wrought me, and Brand the priest." So that this "day's sun-marker," or dial, records that, in the reign of Edward the Confessor, Orm, the son of Gamal, bought the ruined church which was dedicated to S. Gregory, but had fallen down; and he built it anew from the ground in honour of Christ and S. Gregory, who was the pope under whom Augustine was sent from Rome, in 596, to convert the Anglo-Saxons. Moreover, the date of the building is indicated by the fact that it was erected in the time of Earl Tosti. He was a younger brother of Harold, and son of Godwin, Earl of Kent, and on the death of Siward was made Earl of Northumberland in 1056. Being slain 25th Sept. 1066, at Stamford Bridge in a battle with Harold, against whom he had rebelled, the date of the church and dial is limited within ten years, and may be set down as about 1060. Orm, the founder, is named in "Domesday Book" as holding the manor of Chirchebi (Kirkby) under Hugh Fitzbaldric; and there is the further entry, "*Ibi P'b'r et Eccl'a.*" "There is there a priest and a church;" possibly Brand was the "priest" who is named on the dial.

At Great Edstone Church, two miles from Kirkby Moorside, over the south door is another Saxon dial, the lettering on which proves it to be unquestionably of the same age as that at Kirkdale. It is 3 feet 11 inches long, and 1 foot 7½ inches high, being shaped and carved exactly like the other. On the left side, as you face it, there is inscribed, "Lothan me wrohtea," "Lothan made me." Over the semi-circular plane are letters mutilated in the middle, where the gnomon must have formerly been. They seem to spell *orologi — torum*, and may have been *horologium viatorum*—the "time teller of travellers." All we can decipher for certain is the long-forgotten maker's name.

Another Saxon dial is over the south door of the ancient church of Bishopstone, near Newhaven, obviously belonging to the same or even an earlier period. It is inscribed, "✠ EADRIC," possibly the founder's name.

Over the south door of Weaverthorpe Church, near Driffield, Yorkshire, is another dial of like antiquity. It is inscribed, "✠ IN HONORE Sᵀ ANDRE APOSTOL ✠. HEREBERTUS WINTON HOC MONASTERIUM FECIT."

At Headbourne Worthy Church, near Winchester; also in the south wall of the Saxon tower of Barnack Church, near Stamford, and at Swillington, Yorkshire, vestiges of ancient dials may be found. Of later date there is a semi-circular dial at Old Byland, near Helmsley, bearing the maker's name, "TIDEMAN ME FECIT."

The ancient sun-dials in Ireland are supposed to date back as early as the seventh and eighth centuries. The late Mr. G. V. Du Noyer, who closely investigated the ancient relics of Ireland, says that, instead of the specimens he has seen being fixed on churches, as in England, they "occur on flat erect slabs, placed like head-stones in ancient cemeteries." It is the opinion of archæologists that both English and Irish dials of this early period are marked with special reference to the canonical hours of prayer in the Roman Church; just as the dials on Mahometan mosques are intended to remind "the faithful" of their prayers, having frequently upon them a pointer towards Mecca, to which the worshipper must turn his face at his devotions.

In Wm. Leybourne's "Tractates," published in 1682, is an account of a marvellous pyramidical dial, set up by order of Charles II. in the courtyard facing the Banquet House at Whitehall, which stood on a stone pedestal, and rose in six compartments, one less than another, into the form of a pyramid. It was the invention of Francis Hall *alias* Lyne, a Jesuit, and professor of mathematics at Liège. It is said to have contained no less than 271 different dials: some showing the hours according to the Jewish, Babylonian, Italian, and astronomical ways of counting: others making the shadow of the hour lines fall upon the stile, as well as the usual reverse of this; and others displaying things pertaining to astronomy, geography, astrology, &c. The four elements of fire, air, water, and earth, were also represented; and there were portraits on glass of the king, the two queens, (the mother and wife of Charles II.,) the Duke of York, and Prince Rupert. Father Lyne wrote a description, consisting of eleven chapters and seventy-three plates, which is now before us. The cost of this royal toy must have been enormous, for Mr. Timbs says that, "about 1710, William Allingham, a mathematician in Cannon Row, asked £500 to repair this dial: it was last seen by Vertue at Buckingham House."

By the inventor's own showing, the whole construction must have been rather what we may call trumpery; without any grace in the design, and merely fitted to show ingenuity and faulty taste. It could not have been qualified to resist the weather—to which, to be of any use, it must have constantly been exposed—for he complains " that the Diall, for want of a cover, was much endamaged by the snow lying long frozen upon it; and that, unless a cover were provided (of which he saw little hope), another or two such tempestuous winters would utterly deface it." The rough illustrations to the " Brief Explication of the Pyramidicall Diall set up in his Majestey Private Garden at White hall July 24 1669" are quite sufficient to reconcile us to the loss of all traces of this extraordinary conceit.

In " Anecdotes of Painting in England," it is related of the eminent sculptor, Nicholas Stone, under the date 1619, that he made a dial at St. James', the king finding stone and workmanship only, for which he received £6. 13s. 4d. "And in 1622," Stone says, "I made the great diall in the privy garden, at Whitehall, for the which I had £46. And in that year, 1622, I made a diall for my lord Brook, in Holbourn, for the which I had £8. 10s." Also, for Sir John Daves, at Chelsea, he made a dial; and two statues of an old man and a woman, for which he received £7 a-piece.

Of the Privy Garden dial, Edward Gunter, professor of astronomy at Gresham College, seems to have had the setting and arrangement, a description of which he published in 1624, by command of James I. It consisted of a large stone pedestal, with a dial at each of the four corners: there was "a great horizontal concave" in the centre, and dials were also on the four sides. In the reign of Charles II., as Mr. Timbs has noted, this instrument was defaced by a drunken nobleman of the court; on which occurrence Andrew Marvell wrote:—

> " This place for a dial was too unsecure,
> Since a guard and a garden could not defend;
> For so near to the Court they will never endure
> Any witness to show how their time they mis-spend."

In Joseph Moxon's "Tutor to Astronomie and Geographie," published in 1659, there are full directions for making sun-dials of various kinds, and among these he instances, "a solid globe or ball that will show the hour of the day without a gnomon." His instructions are to mark a globe round the equator, with two sets of figures from I to XII; and then erect it, rectified for the latitude, with one XII set to the north and the other to the south. When the sun shines on the globe, the hour is indicated where the shadowed and illuminated parts meet. A fine column or fountain, having four figures upon it, and pouring water from each of the four sides, was placed by Mr. John Leak, at Leadenhall Corner, in the mayoralty of Sir John Dethick, kt., which bore on its top such a globe dial as we have described. A picture of this is given in Chambers' "Book of Days," as copied from Moxon's work.

Covent Garden was originally the Convent Garden belonging to the Abbey of Westminster; and when, cir. 1631, Francis, Earl of Bedford, to whom the property belonged, had the present square formed for the use of a market, it was laid out by Inigo Jones, but not completed. The piazza ran along the whole north and north-east sides, the church of St. Paul was on the west, and on the south was the garden wall of Bedford House, under the overhanging trees of which a few temporary stalls were set up at market times. The square was gravelled over, and in the centre was erected, in 1668, a column, surmounted by a dial with four faces. This we learn from items set down in the churchwardens' accounts of St. Martin.

	£	s.	d.
The Right Honourable the Earle of Bedford, as a gratuity towards the erecting of ye column	20	0	0
Sr Charles Cotterell, Master of the Ceremonys, as a gift, &c.	10	0	0
The Lord Denzill Holles	10	0	0
For drawing a model of the Column		10	0
To Mr Wainwright for the 4 gnomens		8	6

The "Seven Dials" of London was a doric column, standing in the parish of St. Giles-in-the-Fields, where seven narrow streets all converged, and left an open space in the centre of which the pillar stood. It had seven dial faces, each a foot square, which were turned opposite to the respective streets. Evelyn, in his Diary, 5th October, 1694, says, "it was said to be built by Mr. Neale, introducer of the late lotteries, in imitation of those at Venice." Cunningham's "Handbook of London" says, "It was removed in July, 1773, on the supposition that a considerable sum of money was lodged at the base. But the search was ineffectual." The old column is now placed on the green at Weybridge, near Walton-on-Thames, and is surmounted by a ducal coronet, with an inscription to the memory of the Duchess of York. In Gay's " Trivia " we read,

> "Where famed St. Giles's ancient limits spread,
> An in-rail'd column rears its lofty head;
> Here to seven streets seven dials count the day,
> And from each other catch the circling ray:
> How oft the peasant, with enquiring face,
> Bewilder'd trudges on from place to place;
> He dwells on every sign with stupid gaze,
> Enters the narrow alley's doubtful maze,
> Tries every winding court and street in vain,
> And doubles o'er his weary steps again."

In the small garden of Clement's Inn, Strand, there is a life-sized figure of a Moor kneeling at the top of four decreasing stone steps, and holding a sun-dial on his head. Peter Cunningham, in his "Handbook of London," says, "it was brought from Italy, and presented to the Inn by Holles, Earl of Clare, but when or by what Earl no one has told us." There were four Earls of Clare of the Holles family: John, Lord Houghton, created 1624, died 1637; John, succeeded 1637, died 1665; Gilbert, succeeded 1665, died 1689; John, succeeded 1689, created Duke of Newcastle 1694. It was probably the second Earl who was the great builder and improver of the neighbourhood, and erected Clare Market, which is called "a new market," by James Howell in his "Londinopolis," 1657. He says, "There is, towards Drury Lane, a new market called Clare Market;" and the founder of this new market may have been the person who placed the figure with the dial in the little courtyard at Clement's Inn. It is said to be made of bronze, but is covered with black paint.

The account elsewhere given of John Howard's wish to have a sun-dial placed over his grave, which he expressed as he lay on his death bed at Cherson, in the South of Russia, was not fulfilled after his interment. He was buried at the spot he had selected, near the village of Dauphigny, about five versts north of Cherson, and a little eastward from the road to Nicholaif.

His friend, Admiral Priestman, read the Burial Service over his remains; and Dr. Clarke, in his "Travels in Russia" in 1800, says "A monument was afterwards erected over him. This, instead of the sun-dial he requested, consisted of a brick pyramid, or obelisk, surrounded by stone posts and chains. The posts and chains began to disappear before our arrival; and when Mr. Heber made the sketch, the obelisk alone remained in the midst of a bleak and desolate plain."

In Dr. Henderson's "Researches and Travels in Russia," in 1821-2, an account is given of what he justly calls a "cenotaph." This was subsequently built, near the gate of the town of Cherson, beside the Russian cemetery. He says it is of white freestone, about thirty feet in height, surrounded by a wall of the same stone, seven feet high by two hundred in circumference. On the pedestal is a Russian inscription of the following import:—"Howard died 20 January, 1790, aged 65." Towards the summit of the pillar there is a sun-dial; but the only divisions of time exhibited are the hours from ten to two. The sketch of this cenotaph gives a structure in form not unlike a windmill without its sails: the numerals are about one-third from the top. It was built by order of the Emperor Alexander, in honour of the devoted philanthropist, whose last wishes were gracefully remembered; but the sundial was not over the grave. Dr. Clarke could not have seen this building, since Alexander did not begin to reign until 1801.

In the churchyard of Trelleck, Monmouthshire, there is a curious sun-dial that was erected by the vicar in 1689. On three sides of the pedestal are represented, in relief, the three marvels peculiar to the place, viz.:— I. A *tumulus*, supposed to be of Roman origin, and inscribed *Magna mole*, "great in its mound,"—*O quot hic sepulti*, "O, how many buried here!" II. Three stone pillars, whence the name, "Tre-leck," *the town of stones*, and inscribed *Major saxis*, "greater in its stones;" the height of the stones being also given, eight feet, ten feet, fourteen feet, as well as *Hic fuit victor Harald*, "here was Harold victorious." III. Together with a representation of the well and two drinking cups, *maxima fonte*, "greatest in its spring;" and below *Dom. Magd. Probert ostendit*. Trelleck is supposed to have anciently been a large town and place of importance. Tradition states that the pillars were erected by Harold, to commemorate a victory over the Britons; but they are more probably of Druidical origin. The *tumulus* stands in the village, and was once surrounded by a moat, and surmounted by the keep of a castle belonging to the Earls of Clare.

On Chartres Cathedral, under a canopy, is a winged angel, ecclesiastically draped, who is represented as holding a large semicircular dial, dated 1578. On Laon Cathedral is a similar figure. Bolton Abbey has its dial, dated 1646; and at Guiting Grange, Gloucestershire, is a solid square stone dial, dated 1634. At Sudeley Castle (J. C. Dent, Esq.), near Cheltenham, are two dials: one inside the court, and nearly over the entrance archway; the other, much worn, on the battlement above the principal apartments. Both are supposed to be as old as the Castle itself.

The Cross at Chichester, used as a market house, was erected in the fifteenth century by Bishop Edward Story, and repaired in the reign of Charles II. It presents four faces, with dials, to principal streets of the city.

Queen's Cross, in the suburbs of Northampton, has four sides facing the points of the compass, and on each of these is a dial which was fixed in 1712. The tall pillar in the centre of the quadrangle of Corpus Christi College, Oxford, has dials at the top, and was inconvenient in breaking the line of march in the old drilling days of threatened invasion.

One may fairly assume that all our cathedrals have had, in their day, a sun-dial erected upon them, before the clock in many cases displaced it; and the same observation will apply to our parish churches, especially those in country villages, where no other public building exists on which any description of horologe can be placed. And where can the passage of time be more fitly expressed than on the walls of those sacred edifices, in which men periodically meet to be reminded of an approaching eternity? Even at the present day, whenever a new church is built, the first desire of country folk, after placing bells in the tower, is to have a clock face outside. It would be interesting to obtain a note, from all the cathedrals of the country, as to the erection of the sun-dial, and its subsequent removal, if this has taken place.

Two men of great genius in their respective lines, Sir Isaac Newton and Sir Christopher Wren, marked their respect for this teller of time by placing sun-dials at the two Universities where they were educated, as we have already shown: Newton at Queen's College, Cambridge; and Wren at All

Souls' College, Oxford. When Wren designed his great Cathedral of St. Paul, he made the "Clock Room" an important chamber in the vast edifice; but we learn from Charles Knight's "London" that, "on the tower of Old St. Paul's was a goodly dial, made with all the splendour that might be; with its angel pointing to the hour both of the day and night."

In Drake's "Eboracum," B. ii. ch. 2, we read:—" Over the doors of the south entrance of York Minster, by the care of the same Dean (Henry Finch, 1702—1728), was also placed a handsome dial, both horary and solar; on each side of which two images beat the quarters on two small bells." This, as we have elsewhere stated, was removed to give place to a clock; but the clock, in its turn, as we are informed, is likely to be removed.

At St. Mary's, the largest of the Scilly Isles, and near the fort called "Star Castle" (if we remember the spot where we sketched it), is an old cannon stuck upwards in the ground, and over its mouth is a dial plate fixed. What storms must have broken upon it in that tempestuous region!—what hurricanes must have blown around!—what dark nights covered it!—and yet, whenever the sun shines, and cheerfully as if no disturbance ever reached it, the dial face becomes bright again, and the gnomon sends its shadow round the plate.

The Market Cross at Carlisle was erected in 1682, and consists of a Grecian Ionic column, with a plain shaft and pedestal. It rises from the centre of a flight of six circular steps. The column is surmounted with a square block of stone, presenting four faces, which are used as sun-dials: it is crowned with a lion bearing the corporation arms. Above the capital of the column is the inscription, "Joseph Reed, Maior, 1682."

At Clumber, the beautifully-situated mansion of the Duke of Newcastle, in the garden, betwixt the house and a fine marble fountain that was brought from Italy,—with the stone balustrade beyond, and the lake below,—is a pedestal, on which are two iron hoops about a yard in diameter, placed transversely, one inside the other, with a rod across the middle. In the centre of this is a knob, which, when the sun shines, throws its shade on the figures that are marked in gold within the hoops, and thus a very elegant dial is produced.

We have already noticed that in old churchyards the commonest form of sun-dial (see a good specimen in the frontispiece) is that of fixing a plate on the stump of the cross which, by Queen Elizabeth's order, had been cut short for the purpose of arresting mediæval superstition. In Biddulph churchyard, Staffordshire, there is still left a fine mortuary cross entire; and on the face of the square stone, which crowns the well-proportioned shaft, and on which the cross is reared, there is a dial: the whole structure being very interesting as a relic.

The making of dials was formerly reckoned a necessary qualification for masons, when they probably depended less than now on the architect for designs; and they certainly showed both invention and skill. The lineating of the dial plate was a matter of science, to which the highest mathematician of his day sometimes gave his attention. The engraver's art was next called into use; and some existing plates of bronze or copper exhibit great care and beauty of execution. Thus, in the garden of Babworth Hall, Notts (the residence of H. Bridgeman Simpson, Esq.), there is a finely executed plate on a stone pedestal. It was placed there some time ago by the owner, after removal from his estate at Stoke Hall, Derbyshire; and so perfect is the elaborated engraving, showing the months of the year as well as the hours of the day, that even a thought has arisen of turning it into a letter-weight for the library table.

Ring and pocket dials were common in the seventeenth century, and even later. The ring dial, of which an illustration is given, belongs to Mrs. Dent of Sudeley Castle, and was found at Kemerton Court, Gloucestershire. The small piece of projecting brass, with a hole in it, slides in a groove, and acts the part of a gnomon. The small ring is for suspending it in the sun's light, with the side having the sliding hole offered to the sun; whereby a ray falls on the numbers inside the ring, and declares the hour. Such instruments may have been used for astrological purposes.

There is a round pocket dial, with compass, in the Museum at Edinburgh, which was presented by J. Johnstone, Esq.; also a small circular silver dial, which was found in the ruins of an old house at Carnwath.

The Rev. J. Stacye, Governor and Chaplain of the Shrewsbury Hospital, Sheffield, has in his possession a very elaborately worked silver pocket dial, with compass, which is of French manufacture; the maker's name being Sautout Choiz, of Paris. Ring dials used to be made in large quantity at Sheffield.

Amongst the eminent men who have paid respect to the sun-dial may be reckoned George Stephenson, the great railway engineer, who set his son Robert (still a boy at school) the task of making a dial to be placed over their cottage door at West Moor, near Newcastle. Father and son together got a stone, which they hewed, carved, and polished; and, with the aid of "Ferguson's Astronomy," they found out the method of making the necessary calculations to adapt the dial to the latitude of Killingworth. The dial, with the gnomon coming from the sun's face, and dated "August 11th, MDCCCXVI," may still be seen over the entrance to the humble early home of these distinguished men. Let us then venture to hope that the healthy taste of the Stephensons, who by their inventive genius have contributed more than any other men to disturb society in its old stationary customs, may plead in favour of the sun-dial—its preservation and its continued use.

> 'Tis an old dial, dark with many a stain,
>   In summer crown'd with drifting orchard bloom,
> Trick'd in the autumn with the yellow rain,
>   And white in winter like a marble tomb;
> And round about its grey time-eaten brow,
>   Lean letters speak—a worn and shatter'd row—
> "I am a shade; a shadow too art thou:
>   I mark the Time: say, Gossip, dost thou soe?"

In taking leave of the reader, we will quote a line from Dante's *Paradiso*, which is well fitted for a dial motto, and not inappropriate to the action of bidding farewell :—

PENSA, CHE QUESTO DÌ MAI NON RAGGIORNA.

INDEX OF PLACES WHERE THERE ARE SUN-DIALS WITH MOTTOES, AS THEY ARE NUMBERED.

ABBOTS-FORD, 70.
Aberford, 23.
Adel, 333.
Aigle, 188.
Aix-le-Bains, 18.
Albizzola, 35.
Aldeburgh, 115.
Alderley, 326.
Alfrick, 206.
Alghero, 269.
All Souls' College, 149, 218.
Alwalton, 365.
Alzo, 176.
Antibes, 186.
Areley Kings, 17.
Arles, 351.
Arley Hall, 158.
Arenemberg, 329.
Arma, 130.
Arola, 38.
Asti, 258.
"Aunt Judy's Magazine," 118.
Austin Friars, 58.
Aynho, 72, 376.

Bakewall, 125.
Balerna, 252.
Bamborough, 218.
Barmston, 52.
Barnard Castle, 73.

Barnes Hall, 56.
Barton, 366.
Bassenthwaite, 145.
Bececa, 37.
Beverley Minster, 192.
Beziers, 332.
Biella-alta, 128.
Bishopsthorpe, 297.
Bittadon, 163.
Bolton Percy, 191.
Bonneville, 281.
Borden, 178.
Bordighera, 103, 161, 252.
Bowles, Rev. W. L., 65, 237, 300.
Bradbury, 375.
Bradfield, 227.
Brading, 110.
Bramshill Park, 278.
Breage, 373.
Bridlington, 296.
Brittany, 74.
Brockhampton Park, 290.
Bromley, 360.
Brompton-on-Swale, 344.
Bromsgrove, 371.
Brough, 265.
Brougham Hall, 202, 298, 333.
Bruges, 195.
Brussels, 270.
Buxted, 371.
Bywell Abbey, 285.

Cambiano, 66.
Cambo, 59.
Campo Dolcino, near, 37, 115.
Cannes, 45, 148, 204, 341.
Caprile, 43.
Carlisle Castle, 39.
Carenna, 80.
Castasegna, 37.
Castel del Pazzo, near Rome, 267.
Castleton, 111.
Catterick, 88.
Cawder, 115, 363.
Chambery, 61, 77, 196, 256.
Chester, Rev. Greville J., 121.
Chieri, 34, 277.
Cimiés, 242.
Cogoletto, 201.
Coldthorpe, 276.
Colico, 233.
Collaton, 122.
Compton Wynyates, 264.
Connor, diocese of, 279.
Constantinople, 48.
Convento, near to, 327.
Conway, 55.
Coplestone, Bishop, 142.
Corby Castle, 40.
Courmayeur, 10, 19, 28, 79, 137, 139, 354.
Croft, 162.
Crompton, 7.
Cuers, 325.

INDEX OF PLACES.

Danby Hall, 166.
Danby Mill, 189.
Darlington, 257, 271.
Dartmouth, 11.
Dennington, 169.
Derwent, 168, 335.
Dewsbury, 295.
Dolce Acqua, 172.
Dresden, 33.
Dryburgh, 369.
Duncombe Park, 181.
Durham, 280.

Easby, 295.
East Grinstead, 115.
East Harptree, 228.
East Lavant, 122.
East Leake, 190.
East Lodge, 4.
Easton, 371.
Ebberston, 216.
Ecclesfield, 368.
Edinburgh, 16.
Edmond Castle, 274.
Edstone, Great, ix.; 143.
Ellastone, 350.
Elleslie, 24, 84, 187, 193, 218, 235, 260, 338, 346.
Elmsted, 69.
Elsworth, 169.
Ely Cathedral, 138.
Entrèves, 141.
Erith, 237.
Erstfelden, 364.
Esholt Hall, 297.
Eyam, 127.

Farnham Castle, 67, 140, 225.
Farnworth, 115.
Felton, 333.
Florence, 42, 86, 164, 244.
Fornasette, 41.
Fountains Hall, 260.
Fredericton, 124.
Fréjus, 194, 240.
Frome, 115.
Fyning House, 175.

Gale Syke, 214.
Geneva, 361.

Genoa, 247, 249.
Gilling, 88.
Glasgow Cathedral, 75, 216, 324.
Glamis, xx.
Gloucester, 89, 218.
Gourdalou, 131.
Graglia, 2, 69.
Grantham, 181.
Grasse, 13, 59, 204, 282.
Grenoble, 106.
Guildford, 357.
Guilsborough, 83.
Gulval, 295.

Hadleigh, 304.
Halifax, 171, 317.
Hall Place, 134.
Hallstadt, 49, 170.
Handsworth, 296.
Hartest, 335.
Hartlepool, 301, 311.
Hatford, 29.
Hatherley, 179.
Haverfield, 212.
Haydon Bridge, 180.
Hebden Bridge, 236.
Heighington, 63.
Helston, 328.
Hermit Hill, 47.
Hesketh, 9, 313.
Heslington Hall, 26.
Highgate School, 343.
High Lane, 22.
Himbleton, 346.
Hopley, H., 98, 102, 223, 246, 294, 342.
Horton, 221.
Hutton Buscell, 314.
Hurstpierpoint, 345.
Huish Episcopi, 112.
Hyères, 10.

Ingleton, 62.
Isella, 241.
Isleworth, 367.
Italy, 78, 93, 150, 151, 224.

Jamaica, 219.

Karlsbad, 109.
Kedleston, 371.

Kenmare, 126.
Kidderminster, 185.
Kildwick, 218.
Kilnwick, 22, 96.
King's Lynn, 167.
Kiplin Hall, 120, 168.
Kirk Arbory, 117.
Kirk Braddan, 296.
Kirk Michael, 293.
Kirkby Malzeard, 82.
Kirkby Moorside, 338.
Kirkleatham, 73, 374.
Knole Park, 226.

Langen Schwalbach, 53.
Lansdowne Lodge, 126.
La Tour, 25.
Lavagna, 248.
Leadenhall Street, 183.
Leam, 115.
Lesneven, 160.
Leventhorpe Hall, 310.
Lewes, 26, 187.
Leyburn, 312.
Leyland, 229.
Lincoln Cathedral, 31, 218.
Lincoln's Inn, 75, 230.
Louth, 260.
Lower Heyford, 177.
Lugano, near, 259, 261.

Maker, 266.
Malvern, 104.
Marlborough, 62.
Marrington Hall, 305.
Martigny, 76.
Marton-cum-Grafton, 82.
Masserano, 289.
Melsonby, 303.
Menaggio, near, 173.
Mentone, 103.
Middleham, 257.
Middleton Tyas, 157.
Milan, 282.
Millrigg, 46.
Milton, Berks, 216.
Milton, near Gravesend, 316.
Moat Hall, 87.
Monk Fryston, 191.
Monkton Farleigh, 130.
Mont Cenis, 292.

INDEX OF PLACES.

Monthey, 162, 281, 359.
Morden College, 335.
Morvah, 335.
Murano, 13, 147.

New Inn, 309.
New Palace Yard, 57.
Newton House Woods, 135.
Nice, 18, 27, 130, 269, 318, 320, 362.
Northallerton, 209.
Northampton, 5.
North Wingfield, 90.

Oberhofen, 352.
Offchurch, 26.
Offerton, 5.
Ormsby, 333.
Orta, 337.
Ossington Hall, 296.
Over-Peover, 30.

Packwood Hall, 213, 253.
Padua, 340.
Paget, Rev. R. E., 95.
Palermo, 80.
Paris, 153, 287.
Patrick, Isle of Man, 200.
Penrhos, 100.
Perpignan, 197.
Peterborough, 199.
Pickering, 356.
Pilling, 306.
Pino, 269.
Pinsuti, Count, Chateau, 208.
Pisa, 130.
Pocket-dial, 254.
Poirino, 36, 114, 159.
Polesworth, 116.
Pomier, 260.
Pont du Siagne, 131.
Port Royal, 146.
Porto Fino, 6.
Porto Maurizio, 27.
Pra, 348.
Puisseaux, 268.

Redding, Cyrus, 108, 207.
Rho, 165.
Ringenberg, 94.
Riva, 149.
Riviera, 113, 284, 288, 327.

Rome, 133, 241.
Rosenheim, 68.
Rotherham, 122, 218.
Rothwell, near Leeds, 349.
Rougemont, 136.
Rugby, 143.
Rushton, 32, 239.
Rye, 295.

St. Gervais, 3.
St. Hilary, 8.
St. Just, 260.
St. Madeons, 308.
St. Pierre, 107, 330.
Salisbury, 144.
Salzburg, 286.
San Michele, 53.
San Remo, 284.
Sandal, 353.
Sandhurst, 75.
Schwyz, 44.
Seaham, 302.
Sedbury Hall, 67.
Sestri Levante, 263.
Sestri Ponente, 1.
Shaftesbury, 335.
Sheffield, 346.
Shenstone, 122.
Shutts, 51.
Siagne, Bridge over, 131.
Sleights, 335.
Sleningford, 101.
Smeaton, Great, 143, 231.
Sordevole, 217, 252.
South Stoneham, 273.
Southall, 272.
Spain, 245.
Spotorno, 321.
Sprawley, 5.
Sproughton, 355.
Staindrop, 155.
Standish, 174.
Stanhope, 333.
Stanwardine Hall, 129.
Stassano, 331.
Stazzano, 60.
Stirling, 119.
Stoke Newington, 274.
Stra, 154.
Strevi, 251.
Superga, 80.
Switzerland, 319.

Temple, 22, 57, 218, 255, 309, 344.
Thornton, 306.
Threckingham, 262.
Thun, 50.
Torrington, 99.
Tours, 14.
Trafiume, 243, 334.
Tunbridge Wells, 377.
Tuscany, 132.
Tutbury, 54.

Uppingham, 123, 182.
Urugne, 363.
Uttoxeter, 105.
Val Sesia, 210.
Vallauris, 91.
Varallo, 339.
Varenna, 222, 233.
Venice, 115.
Versailles, 250.
Vian, 291.
Vignale, 220.
Visp, 205, 211.
Voltri, 97.

Wadsley, 203.
Walgrave, 372.
Warrington, 198.
Watton Abbey, 315.
Wellingborough, 275, 234.
Welwyn, 12, 67.
Wensley, 15.
Wentworth, 152, 325, 358.
Wentworth Woodhouse, 358.
West Felton, 296.
West Ham, 238.
Whitby, 215.
Whitley Hall, 187.
Willerby, 307.
Willesden, 62.
Wilton Bridge, 71.
Winchester, 20, 62, 336.
Wolsingham, 64.
Woodhouse Eaves, 30.
Woolwich, 20.
Wordsworth's "Excursion," 311.
Wortley, 47, 56.
Wragby, 370.
Wycliffe, 155.

Yaxley, 221.

REMARKABLE SUN-DIALS WITHOUT MOTTOES,
AS THEY ARE PAGED.

AHAZ, dial of, xi.
Babworth-Hall, 151.
Banquet-House, Whitehall, 144.
Barnack, 144.
Biddulph, 150.
Bishopstone, xix. 143.
Bolton Abbey, 148.
Boroughbridge, xix.
Bowland, 140.
Buen Retiro, 138.

Carberry Tower, 140.
Carlisle, 150.
Cawston, 140.
Chartres Cathedral, 149.
Chelsea, 145.
Chichester Cross, 148.
Chinese, xviii.
Clement's Inn, 147.
Clermont-Ferrard, 138.
Clumber, 150.
Corpus Christi College, 149.
Covent Garden, 146.
Craigmillar, 140.
Cramond House, 140.

Dial of box edging, xxi.
Dundas Castle, 140.

Egyptian dials, xiii.
Elmley Castle, 141.

Floral dial, xxii.
Florence, 139.

Glamis Castle, xx.
Great Edstone, xix. 143.
Grecian dials, xiii.
Greek-Egyptian dial, xiv. 137.
Guiting Grange, 148.
Gun dial, xviii.

Headbourne Worthy, 143.
Heriot's Hospital, 140.
Holborn, 145.
Holyrood, xix. 139.
Howard's dial, xxii. 147.

Inch House, 140.
Indian dials, xiii.
Innes House, 140.
Irish dials, 144.

Jacques' dial, xxi.
Japanese dials, xviii.

Kelburne House, 140.
Kirkdale, xix. 142.

Laon Cathedral, 148.
Leadenhall Corner, 146.
Leuchars, 140.

Madely Hall, 137.
Mahometan dials, xvii. 144.
Meggetland, 140.
Melrose Abbey, 141.
Melville House, xix.
Midmar Castle, 140.

Northampton, Queen's Cross, 149.
Northbar, 141.

Old Byland, 144.

Peruvian Indian dials, xvii.
Pinkie House, 141.
Pocket dials, 151.
Polton, 141.
Privy Garden, Whitehall, 145.
Pyrenees Peasant's dial, xxi.

Queen's College, Cambridge, 149.

Ring dials, xxi. 151.
Roman dials, xiv.
Rubislaw Den, 140.

Saint James', 145.
Saint Paul's Cathedral, 149.
Scilly Isles, 150.
Seven Dials, 146.
Silver dials, 151.
Sudeley Castle, 148.
Swillington, 144.

Temple, Sir W. his dial, xxii.
Trajan's dial, xvii.
Trelleck, 148.

Weaverthorpe, 143.
West Moor, 151.
Woodhouselee, 140.

York Minster, xx. 149.

PRINTED BY WHITTINGHAM AND WILKINS, TOOKS COURT, CHANCERY LANE.

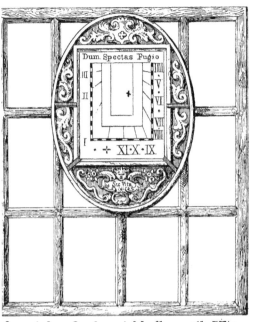

In a window of a shop at Marlborough, Wilts:
Nº 62.

Sun dial at Sedbury
Eheu fugaces! Nº 67.

Catterick Church
Nº 88.

Haydon Bridge Northumberland,
over the South door of the Church. Nº 180

Rougemont.
Canton de Vaud.
Switzerland.

Je luis pour tout le monde
Ton ombre passe avec vitesse et la fin approche.
vec Rapidité O mortel.
N° 136

Vides Horam et Nescis Futurum.

Pra Riviera di Ponenti
N° 348.

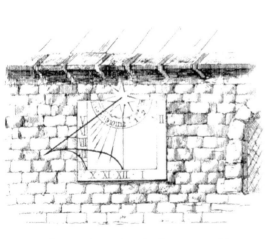

Gedenke dass du sterben musst!
Ringenberg near Interlachen.
N° 94

ΓΑΡ ΝΥΞ ΕΡΧΕΤΑΙ.

Abbotsford N° 70.

Dials at Glamis Castle. NB.
Page XX.

Dials at Kelburne House NB
Page 140.

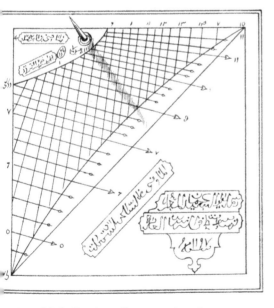

Turkish Dial at Constantinople.
Nº 48.

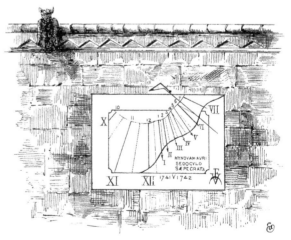

Court of the Mairie Perpignan.
Nº 197.

IRREVOCABILIS HORA
CANNES. On a house at the entrance of a bridge Pont du Siagne
GOURDALON on a shed 2 miles from Cannes.
N° 131.

At Rosenheim
N° 68

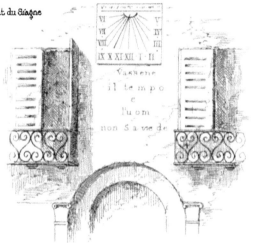

Via Brondolo Padua
N° 340.

Rue de Franco, Nice   N° 130.

A RING DIAL of the XVIIth cent[y] found at Kemerton Court Gloucestershire now in the possession of M[rs] Dent.

Page 150

FOR THE LADY ABNEY
AT NEWINGTON.
1735.
So Rolls the Sun So Wears the Day
And measures out Life's painfull Way
Thro' shifting Scenes of Shade and Light
To Endless Day or Endless Night.

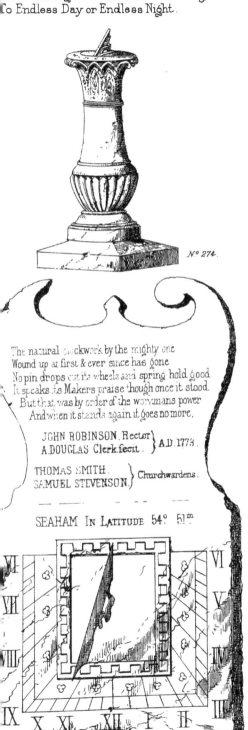

N° 274.

The natural clockwork by the mighty one
Wound up at first & ever since has gone
No pin drops out its wheels and spring hold good
It speaks its Makers praise though once it stood.
But that was by order of the workmans power
And when it stands again it goes no more.

JOHN ROBINSON, Rector } A.D. 1773.
A. DOUGLAS Clerk fecit.

THOMAS SMITH } Churchwardens.
SAMUEL STEVENSON.

SEAHAM In Latitude 54° 51ᵐ

N° 302.

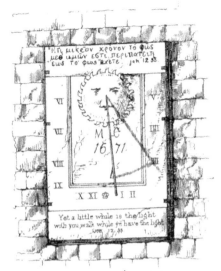

School-house at Ayhno.
N° 72.

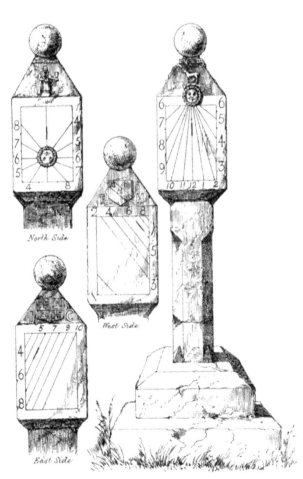

Dial at Dryburg Abbey. N° 369.

For EU product safety concerns, contact us at Calle de José Abascal, 56–1°,
28003 Madrid, Spain or eugpsr@cambridge.org.

 www.ingramcontent.com/pod-product-compliance
Ingram Content Group UK Ltd.
Pitfield, Milton Keynes, MK11 3LW, UK
UKHW030901150625
459647UK00021B/2680